# Whiskey Whiskey Papa

chronicling the exciting life and times
of a pilot's pilot.

Norman Avery

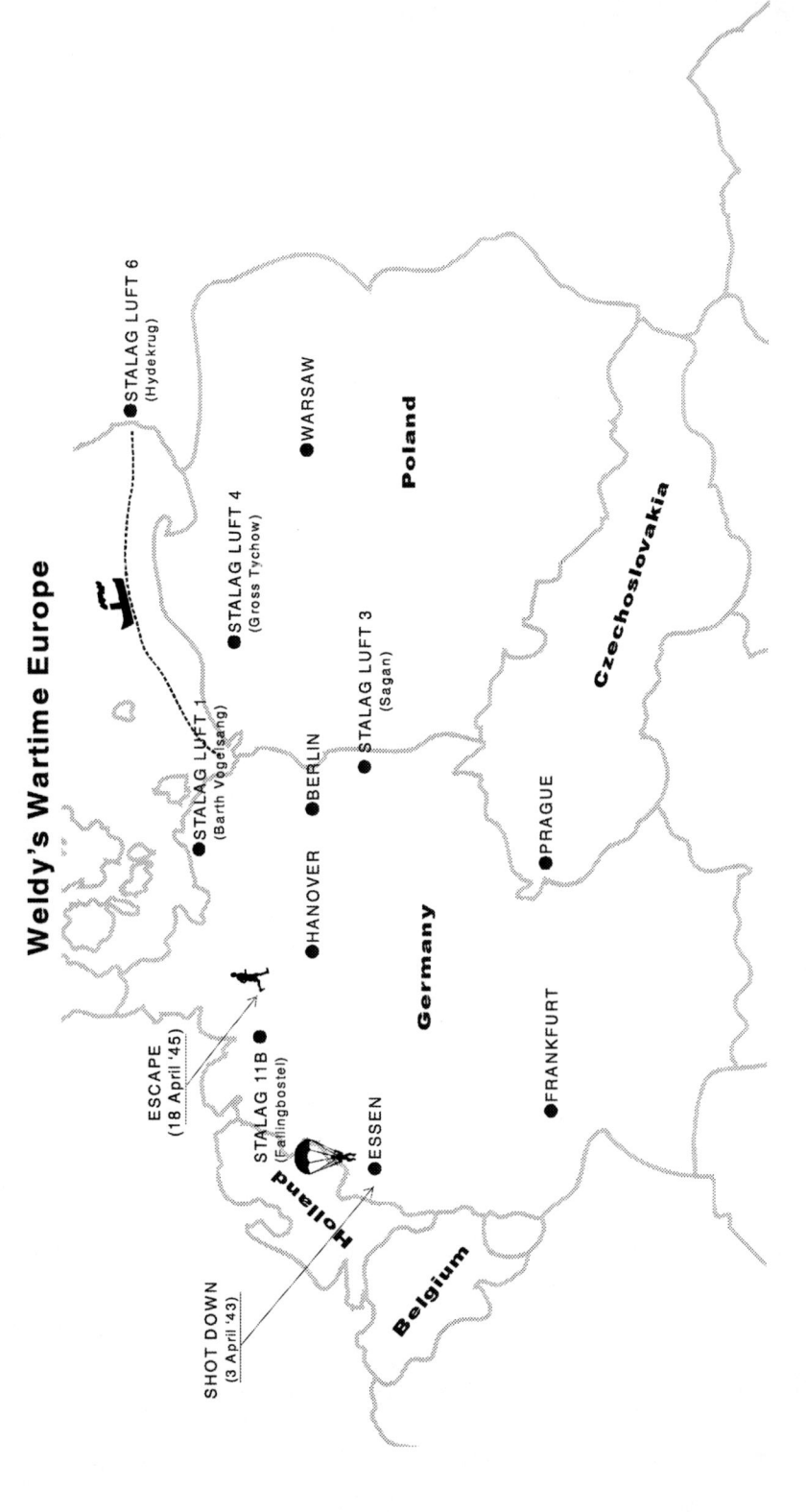

Copyright © Norman Avery, 1998

All rights reserved. No parts of this book may be reproduced, stored in a retrieval system, or transmitted, in any form or by any means, electronic, mechanical, recording, or otherwise without the prior written permission of the author, except for purposes of a review, in which passages may be quoted in print, or broadcast on television or radio.

Canadian Cataloguing in Publication Data
Avery, Norman Ray 1929 —

**WHISKEY WHISKEY PAPA :**
chronicling the exciting life and times
of a pilot's pilot.

ISBN 0-9684211-0-5

1. Phipps, Welland W. 2. Air pilots — Canada — Biography.
3. Aeronautics, Commercial — Canada, Northern — History.
4. Title.

TL 349 P445A94 1998 387.7'092 C98-900952-1

Cover design and illustration by
Jim Turner of Pixelgraph Studio, Ottawa.

Published By
Norman Avery
1057 Pinewood Crescent,
Ottawa ON
K2B 5Y3
(613) 820-3239
Internet: navery@cyberus.ca

Composed by the author on an Atari Mega4 STE,
converted to MS Word and designed in MS Publisher by PixelGraph Studio
printed and bound in Canada by M.O.M. Printing

# Table of Contents

## Chapter One: Throttle set and contact
From a five-year-old playing with explosives through the horrors of war and back again .................................................. 7

## Chapter Two: Picking up the pieces
From sprog pilot with a penchant for backyard barnstorming through to a commercial career in the bush ...... 33

## Chapter Three: Reshaping the aviation scene
From the quiet idea through to the brutal reinvention of the airplane and some of the fun involved ..................... 49

## Chapter Four: Bradley Air Services
From the birth of Tundra Tires through the opening of the High Arctic to practical exploration. ......................... 74

## Chapter Five: The rebirth of Atlas
From the move to Resolute through to view of the north from a political soap box ............................................................ 94

## Chapter Six: That southern yearning
From wind-down of the Arctic experience to a new life ashore and afloat in a warmer atmosphere ...................... 120

## Chapter Seven: Remembering their father
Six children tell of the challenges their father gave them in the air, on the ice and at sea ............................................. 128

## Appendix: Honours and awards
From a narwhal tusk with Inuit signatures to the Aviation Hall of fame and the Order of Canada. ....... 151

## A final word:
A colleague outlines the qualities of a man who would have been admired by C.D. Howe ........................................ 161

**Remembering Len McMann and Eddie Thomas
whose wings were clipped too soon.**

**Norman Avery's** interest in aviation goes back some 60 years when he saved corn flakes box tops and corn syrup labels to redeem for aircraft pictures. When civil aviation resumed after the war he landed a job pumping avgas into the tanks of a bush Norseman and pumping water out of its floats. His pay: flying lessons leading to a licence endorsed to fly on floats skis and wheels.

After a stint in the RCAF he became a journalist. He met Weldy Phipps while writing a weekly aviation column in The Ottawa Citizen in the sixties. But even a reporter's professional prying was not enough to extract much personal information from the enigmatic Mr. Phipps. And so, after his death in 1996, it remained for his family, friends and admirers to fill in the blanks of his remarkable life as recorded in Whiskey Whiskey Papa.

The author has a feeling that he has only uncovered the tip of the iceberg.

Welland W. (Weldy) Phipps OC
1922 — 1996

Drawing from the gallery of the Aviation Hall of Fame,
courtesy of the artist, Irma Coucill

*Self discovery comes when man measures himself
against an obstacle*

– Antoine de Saint Exupery

# WHISKEY WHISKEY PAPA

## Chronicling the exciting life and times of a pilot's pilot

I am tempted to use the term legendary to describe Welland Wilfred Phipps. But my dictionary says a legend is a traditional story popularly regarded as historical myth and, among other things, "...the story of the life of a saint". Weldy's entitlement to sainthood might call for some adjudication, but there can be no challenge to his place in aviation history. Citations in his honour range from syllabic signatures on a narwhal tusk by appreciative Inuit friends to membership in the Order of Canada. And, there is the Trans-Canada McKee Trophy, The Order of Silver Wings, The Order of Polaris, The Order of Flight, and membership in the Aviation Hall of Fame.

I first met Weldy during the sixties when I wrote an aviation column at The Ottawa Citizen. This led me into the local aviation fraternity where I quickly learned that this man was no ordinary flyer. But learning of all his accomplishments was not a simple task. He was not one to recount his experiences willingly, preferring to forge ahead rather than rest on his laurels.

He formed many friendships in the wartime Air Force, in prison camps, and in the hangars and cockpits of civil aviation. The memories of these friends have added a colour to his life that he would decline. Some tales may have ripened in the retelling. Various writers, caught up in the excitement, could easily appropriate a little poetic licence. I hope that I am not among them.

I am particularly indebted to Fran Phipps for her characteristic enthusiasm for (nearly) everything her husband undertook. Her own contributions to Weldy's prominence will never be adequately rewarded. And so, as you marvel at his accomplishments, please credit Fran for a great deal of his success. I also want to thank Geraldine Phipps Trudel, Weldy's sister, for her help in filling in many details of his life. The scrap book she so lovingly maintained from his earliest days was made available to me, and our conversations brought to mind many more facts. The Phipps children submitted their favourite memories of their father. Their contributions in total spanned both the flying and the sailing days, and provided a glimpse into the basic nature of a man driven by challenge.

Readers will note the occasional reference to miles and kilometres as well as Eskimo and Inuit. These terms relate to conventional usage at the time of the incidents being reported.

# *Appreciation*

I have spoken to so many people about Weldy that I might miss some in expressing my gratitude. And so, fingers crossed, I wish to thank:

**Hugh Clee**, Ocean Park, B.C., Air Force buddy and fellow POW; **Judge Edward Houston**, Ottawa, fellow POW, later legal counsel; **Angus Morrison**, Almonte, Ont., Atlas Aviation Ltd. No. 1., Weldy's first employer; **Bill Law**, Merrickville, Ont., fellow barnstormer, mechanical engineer, Spartan Air Services, deHavilland Canada; **Bill Doherty**, Ottawa, Spartan maintenance engineer & Atlas Aviation supporter; **Bob Bolivar**, Ottawa, Spartan navigator; **Dalton Muir**, Ottawa, wildlife biologist & filmmaker; **Bill Peppler**, Ottawa, Spartan pilot, later COPA exec.dir.; **Ray Lachance**, Ottawa, Spartan navigator; **John Harley Deacon**, Kinburn, Ont., Spartan engineer; **Jim Lotz**, Halifax, geographer and Arctic explorer; **Al Macnutt**, Abbotsford, B.C., Spartan pilot; **Jim Murray**, Calgary, Spartan radio operator; **Jean Greenfield**, Ottawa, Spartan operations room manager; **Tony Little**, Nepean, Ont., president, POW Association; **Ray Stone**, Ottawa, freelance radio reporter; **Tom Appleton**, Survair and deHavilland pilot; **Bob Fowler**, Weston, Ont., Spartan pilot, later deHavilland test pilot; **Robert J. Kerr**, Toronto, co-founder, IMAX Corporation; **Stu Green**, Ottawa, private pilot, Phipps student; **Russ Hall**, Ottawa, Spartan executive; **Joe Holmes**, Ottawa, fellow high school student and Weldy's first passenger; **Vic Johnson**, Ottawa, editor, Airforce magazine; **Ken McLennan**, Deseronto, Ont., Bradley pilot; **Walt Petersen**, Saint Paul, Minn.., engineer, Plaisted Polar Expedition (1968); **Doug MacLeod**, Red Deer. Alta., Spartan mechanic. For their guidance in the complexities of research, editing and publishing I must include **Jim Turner** and **Bernadette Quade**, Aylmer, Quebec, **Dale MacMurray**, Ottawa; **Elward Burnside**, Toronto; and for patient access to their resources, the staffs of the **National Archives of Canada**, the **National Aviation Museum** library, **The Aviation Hall of Fame**.

(affiliations noted relate to positions held during the Weldy days)

~~~~~~

### The Fortune Cookie...

One stormy evening Weldy and I headed out from Toronto to Ottawa in Piper Apache – CF-NME. We had been into Downsview airport to check on the progress of Whiskey Whiskey Papa, Twin Otter No. 12, which was under construction by deHavilland for Atlas Aviation Ltd., the company Weldy established at Resolute Bay, N.W.T.

It had become apparent earlier that afternoon that something was wrong with NME's radio: sometimes it worked and sometimes it didn't. The malfunction set Weldy's analytical brain to work on the cause. He decided that when the wheels were put in the down position the radio either did not transmit or did not receive. Pull the wheels up and communication with controllers was restored. Put them down to land and all outward calls went unanswered. We were cleared final and approached the runway with the wheels up. On close final the wheels were lowered and we landed. Safely on the ground, two-way communications worked again.

Weldy had an inkling that the landing had not solved the problem. On the return flight, which had included a shopping trip into Malton airport, he conducted a test by calling Toronto tower again – wheels down, wheels up. Sure enough, the snag was the same as before. The two-hour trip ahead of us offered him an opportunity to make some airborne repairs. As I handled the right seat, circling over Toronto, Weldy took a flashlight and wormed his way down under the panel, head first. His feet kicked the control column and his shoulder bumped the rudder pedals. I deflected his manoeuvring kicks from hitting the panel switches. Through this confusion I raised and lowered the gear while he fiddled with wires.

He had taken the microphone down under the panel and proceeded to block the airwaves with his hesitant style of transmission. "Aaahhhh (lengthy pause) Toronto tower – aahhh (lengthy pause) this is... (pause) (pause) November – aahh – Mike – aahhh – Echo...". And so on.

I could only imagine the exasperation of the controllers and other pilots wanting to use the frequency as Weldy carried out tests in the same protracted speech. But, eureka! Full service was restored. Meanwhile, as we circled the east end of Toronto, the weather deteriorated rapidly. We were buffeted by strong winds under a ceiling that descended over us like a dark grey tent.

Weldy, who had been under the panel for about fifteen minutes, resurfaced. His feet thrashed around, kicking me as he tried to wriggle his body back into a more upright position, and I tried to counteract his interference with the controls. He was not quite all the way back when he grabbed the wheel and cranked the Apache into a split-ass turn. Even with his reduced night vision from the flashlight, he had spotted another aircraft coming right at us as it tried to duck under the lowered ceiling. Unaffected, he completed his awkward return to the seat and reattached his seat belt. He gave no sign of concern for my obvious lack of vigilance; his satisfaction in having solved the radio problem seemed to override the near-miss incident.

But that was not the end of our troubles. Lightning began to flash around us. The turbulence was far beyond the comfort zone, and only brief glimpses of the ground showed between low scudding clouds. Weldy took it all in stride. He said, quite casually, "I think there's a beacon at Oshawa. Maybe we can get in there."

He set about tuning the automatic direction finder and listening for a signal that would identify Oshawa airport. The needle on the ADF was swinging wildly from the charged atmosphere, and it was too rough to pull out the Canada Air Pilot to check the procedure necessary for a safe letdown into Oshawa. But Weldy, still laid back comfortably, seemed confident that he had located the radio beacon. He made the essential turns on instruments alone and proceeded to wrestle the Apache down through the clag. As we descended blindly, I looked anxiously for some sign of airport lights. None appeared. And then we broke out of the line squall right over the runway numbers. I had worked up a sweat but Weldy was ho-hum cool.

We stayed overnight in Oshawa where we dined in a Chinese restaurant. I cracked open my fortune cookie and read some proverb about the weather. We had a laugh, and Weldy asked, "Where did you get that?" When I told him it came out of the cookie, he said, "I think I ate mine."

The airmanship and technical skill displayed on that flight were vintage Phipps. His career in the High Arctic saw many more challenging diagnoses to be followed by on-the-spot solutions: one damaged undercarriage fixed with ancient bed frame parts, a rudder held in place with two ski poles, a tailwheel replaced with a shovel, a fuel pump restarted with a hammer, an oil leak on a prop resealed with a piece of a corn flakes box, and a tail section made flyable with a sheet of plywood.

He was asked once about the risks he took flying single-engine aircraft in the High Arctic. What would happen if he had an engine failure over the open ocean?

"I'd stick the nose straight down," he replied, "and make a splash you could see for miles."

"Like hell you would," said a bystander. "You'd climb out and fix it."

—N.A.

~~~~~~

# Chapter one
## Throttle set and contact...

For a man who accomplished so much in his 74 years, Weldy Phipps got anything but a jump-start on life. He was a "sickly" baby, a victim of severe colic. He could not tolerate any food and so went into a frightening decline. The attending doctor tried everything, finally advising Mrs. Phipps that the best thing she could do for her son was to have him baptized. Resigned to the baby's fate, she took the doctor's advice. Very shortly afterwards, he began to gain normal health.

He was baptized Welland, not exactly a trendy name of the time. But it actually had a link to the Welland Canal. A senior engineer on one of the canal reconstructions had named a son Welland. That child had an English nanny. The English nanny was a friend of Nellie Phipps. Mrs. Phipps liked the name Welland. The second name was for his Uncle Wilfred. But in the registration of the name, Wilfred got translated to Wilfrid, apparently by a clerk who was more familiar with the French version. Official measures to translate it back to Wilfred were never taken although the English version was the only one ever used.

The label 'Weldy' seems to have gained coinage after the war. It was in use by his fellow students when he returned to high school in 1945. But even the nickname created confusion. In many accounts of his accomplishments, writers have tried to guess what 'Weldy' stood for. Top of the list has been Weldon. It appeared as Weldom in one publication. And when he won the Around PEI sailing race in 1980 the Summerside paper gushed generously about Welby. At that time he had lived on the island for five years. At home he was Welland, a fact his father would re-establish firmly any time anyone called him Weldy.

The official record of his first wartime operational flight lists him as Phibbs. His air force buddy Hugh Clee called him 'Cub'. The name that stuck throughout his civilian aviation career, however, was Weldy. And when calls for "Whiskey Whiskey Papa" (his initials and the phonetics for his Twin Otter WWP) went out across the High Arctic from his radio base at Resolute Bay, that sobriquet became synonymous with the man.

## We have blast-off...

Welland Phipps demonstrated his fascination with challenge at an early age. It was a characteristic that served him well throughout his life, only to be threatened from time to time by misadventure. One of his experiments took place at about age five. His father had been blasting rock to excavate for an addition to their Ottawa home. A cautious man, he had placed the blasting caps high in a cupboard out of reach of mischievous hands.

To young Welland this offered an irresistible challenge. The shiny metal devices, about the size of a two-inch pencil, looked to him like miniature firecrackers. He knew they went BOOM. Unobserved, he climbed up to the top shelf of the cupboard and removed some of the detonators. Then, with a slightly older friend, he set about trying to explode one. But the metal casing proved hard to crack. Smashing it with a rock didn't work, and so they turned to the railway for help. They placed one on the tracks and waited patiently for the train to set it off. But the vibration jarred the cap off the rail before the train could crush it. They returned to the rock idea. After a few smashes with a bigger rock it finally exploded, driving particles of shrapnel into the legs of the friend. Welland was not hurt.

Following a vigorous discussion between fathers, Mr. Phipps took Welland by the hand and walked all around the neighbourhood "looking for a policeman" to report the crime. They didn't find one but Welland got the message.

~~~~~~

## Head conditioning...

A message of another sort was delivered to him by his Grade One teacher in 1928. His sister Geraldine recalls seeing some excitement ahead of her as she walked towards school. When she reached a group of children gathered on Ottawa's Richmond Road she was told: "Welland's been hit by a car." He had been rushed to a nearby doctor's office. Gerry stopped long enough to pick up a pocketful of her brother's alleys that had been strewn on the road by the impact, and continued to school.

When she returned home for lunch she asked her mother how Welland was. "Why?" her mother asked. When told that he had been hit by a car, Mrs. Phipps ran in search of her son. The doctor had stitched a cut in the top of his head. His leather aviator's helmet had spared him more serious injury.

Mrs. Phipps was highly annoyed to find that he had been delivered to the school instead of home. The teacher was less impressed with the injury, saying: "Maybe this will smarten him up."

~~~~~~

## Hot wheels with traction...

Welland seems to have inherited mechanical skills from his father. Albert Phipps could build or fix anything. He was also an avid gardener who won many prizes for his flowers. His outstanding tulips earned him an ever-increasing supply of bulbs, thereby establishing a continuous cycle which ultimately produced 7000 annual springtime blooms. The secret of his success was in the good black soil in his garden. It did not originate in the garden but could be extracted from a rich peat bog that was located in the heart of nearby Tunney's Pasture, long since supplanted by a vast federal government complex.

To dig up and move the black earth to his garden, Albert Phipps required a truck. He had a Star automobile, circa 1932, in which he saw the solution. He simply hacked off the back end and installed a box which transformed the Star into a truck. The vehicle fascinated his two sons. As a pre-teenager, Weldy was allowed to drive the truck through the field to the peat bog and back with a load. But shovelling was not as much fun as roaring around over forbidden territory. His younger brother, Gordon, was a willing passenger. When father Phipps was out of sight, the Star made thrilling trips along back roads which led to the Ottawa River. But back roads in those days (now paved Ottawa streets) left a bit to be desired.

On one of the secret trips the Star got stuck in the soft ground. Gordon was instructed to push the vehicle while Welland gunned the engine. All was looking good until Gordon slipped on the wet soil and fell under the spinning wheel. He provided the required traction as the truck ran over him and out of the mire. The light weight of the vehicle and the soft ground under the wheels saved Gordon from any worse injury than a scare.

~~~~~~

## Hotter than a Harley...

As a teenager, Welland's mechanical challenges intensified. He no longer considered the lowly bicycle an adequate vehicle. He aspired more to a motorcycle. The compromise, as he saw it, was to mount a small engine on the carrier over the rear wheel of his bike. That was not a difficult proposition to such a mechanically talented boy. And so a small air-cooled motor was found, mounted and coupled to the rear sprocket of the bike. When all work was completed the engine was fired up and readied for its maiden voyage.

To celebrate the accomplishment, Welland invited Gordon to go along as his first passenger. The motor was revved up. The ersatz motorcycle took off down the street in a cloud of smoke. But it was a short trip. The not-so-easy riders dismounted in a hurry when the gas tank burst into flames and set their pants on fire.

~~~~~~

## In quest of wings...

Welland Phipps was a restless young man. He had little interest in the academic challenges of high school, preferring instead to reach out into the real world for challenge and excitement. And so, at the end of his second year at Ottawa Technical High School he left the stifling boredom of the classroom. The Second World War had started, stirring the imagination of many teenagers. But when dreams clashed with reality it was a sobering experience.

Welland's departure from school proved to be an important milestone in his education. He learned very quickly that hand-feeding a printing press was about as unchallenging and boring as work could get. It brought into clear focus how valuable an education could be. But he was not one to let mistakes stand in the way of his ambitions for very long.

He had served his apprenticeship as a model aircraft builder. And he was a frequent visitor to the Ottawa airport to watch the increasing air force traffic. Neighbours not much older than he were enlisting. He could hardly wait to join them. He outlined to an RCAF recruiting officer his lifelong interest in airplanes, his affection for things mechanical and his desire to become a pilot. All he got from the recruiter for his enthusiasm was a pat on his 17-year-old head. But his pitch, accompanied by a recommendation from his employer, must have left a favourable impression.

The letter from the Photogelatine Engraving Company, Canada's home for souvenirs and novelties, stated:

> *This is to recommend W. Phipps. He is both honest and industrious and a good all around worker. He has given every satisfaction during the time he has been in our employ.*
>
> *H. Guile, superintendent*

He had been turned back on his first attempt to enlist, but the Air Force had not lost track of his enthusiastic application. Just after his 18th birthday, his mother received a call from the recruiting office. Welland was to report to the Air Force right

## Whiskey Whiskey Papa

away, if he was still interested in joining. But he was out of town, visiting somebody his mother did not know somewhere around the Ottawa Valley town of Douglas. Mrs. Phipps called the telephone operator in Douglas and explained the importance of her search. The operator, taking advantage of the rural party line, was able to locate Welland. She got him to a phone and allowed the long distance call 'on the house'.

Welland was sworn in on July 30, 1940, as R54222 AC2 Phipps W.W. He longed for the white flash of an aircrew trainee in his wedge cap and he could almost see pilot's wings over his left breast pocket. But aircrew training required a higher level of education, a point emphasized by the rude slash of the recruiting officer's pen through the "pilot" checkpoint the recruit had marked in the enlistment document. His aptitude for mechanics, however, won him acceptance for technical training. But first came basic training at No. 2 Manning Depot in Brandon, Manitoba. He had to learned to march, take orders and polish boots and buttons.

After four weeks in Brandon he was posted to No. 1 Technical Training School at St. Thomas, Ontario for five months' training as an aero-engine mechanic. Upon graduation he was moved to his first staff job at No. 7 Service Flying Training School in Macleod, Alberta. The training environment was under rigid control, but life on the flight line for a teenager was something else. Young Phipps became the object of torment by older, more experienced airmen. He suffered the usual pranks – 'Phipps, go down to Hangar 2 and get two pails of propwash!' That meant walking down the flight line and struggling back with two buckets of lead shot used as weights by parachute packers. Or it might be: 'Phipps, help me test this airspeed indicator. Blow on the pitot head.' The instrument mechanic would have flipped the switch to heat the pitot head and burn the greenhorn's lips.

When the situation became unbearable for AC2 Phipps, an older mechanic came to his rescue. Frank Martin, a tough Saskatchewan farmer six years older than Welland, put a stop to the harassment. The two became lifelong friends. They were posted overseas Feb. 3, 1942, but separated, Frank going to North Africa and Welland to 409 Squadron in England where he went to work servicing the engines of Beaufighters.

On 409 Squadron, Leading Aircraftman (LAC) Phipps met Hugh Clee, a British Columbian and fellow mechanic. The two linked up at work and play. H.A.M. Clee, or "Hambone" as he became known, was a bit older. He named his pal "Cub" and became his mentor and soul mate. Hambone teased Cub a lot – about girl friends, and particularly, his lack of decorum. Air Force regulations stated that "Caps - Field Service - were to be worn on the right hand side of the head so the lower button is in the centre of the forehead and the edge of the cap is one inch above the centre of the right eyebrow." Hugh Clee was always very well turned out while young Phipps preferred a deportment a few notches below the spit-and-polish military standard. "Sometimes," Hugh said, "I'd tell Cub to walk behind me as if we weren't together because of the way he wore his uniform." The precise placement of the cap - field service - was an order just asking to be violated as far as LAC Phipps was concerned.

Overseas they worked on the doorstep of the shooting war but out of the real action. But dreams of being pilots drifted very close to the surface at all times for the two mechanics. And then, after they had resigned themselves to the full war as grease monkeys, the opportunity arose: an opening for aircrew recruits was posted. There were slots for pilots, navigators and flight engineers, to be selected from among the groundcrews if they could qualify. In the activity that followed, both applied to remuster as pilots. But now it was a case of over-qualification. Their technical training and experience suited them best for transfer to flight engineers' school. And so the two were sent to the RAF's No. 2 School of Technical Training at Athens, Wales.

Flight engineers' training included every possible technical aspect of an aircraft's hull and engines. It further extended into the area of radio operations, Morse code, meteorology, navigation (including astro-navigation), recognition of friendly and enemy aircraft, the geography of war zones, and all aspects of gunnery and armaments. In other words, the FE had to be ready to do everybody else's job on the aircraft.

In September, 1942, both received their single 'E' wing and promotion to sergeant. Flight engineer was a relatively rare position for Canadians, the majority hitherto having come from the Royal Air Force. Both were posted to 405 City of Vancouver Squadron based at Beaulieu, Hampshire.

Early in the briefing speech that greeted flight engineers arriving on 405 Squadron was the fact that there were three pubs "within crawling distance of the Mess." What was excluded was the fact that the huts that were to be their living quarters were drafty, cold and most uncomfortable for Canadians raised on central heating in insulated homes. The "bog" or washrooms were far from the quarters.

New arrivals were advised to go to station stores and draw a watch, Flight Engineers' Notes, a knife, tool kit, vacuum flask and torch. For transportation each was entitled to a bicycle. (The archives documents show no warning about the ground hazards of aircrew life. The medical list for one month reports seven men treated for bicycle accidents and only six for venereal disease.) As the seriousness of the briefing continued FEs were reminded that they must also know FE "gen" (for general information), bomb aiming, bombsights, icing conditions, turrets and the operation of same.

No. 405 Squadron was formed in 4 Group April 23, 1941, at Driffield, Yorkshire. Later it was adopted by the city of Vancouver, an occasion marked by the donation of 10,000 cigarettes by the ladies of that city. It had the distinction of being the first Canadian bomber squadron on duty. Its first taste of action came about six weeks later on 4 June, 1941, when one of its Wellingtons was attacked in the air. After six months of bomber operations, in which the squadron lost 33 crews, the unit was posted to Coastal Command. The new role was considered by the crews a good rest. They had a ditty that went:

*It's foolish but its fun*
*And if you want to stay alive*
*Get the hell off 405.*


The change to Coastal Command coincided with Sergeant Phipps's introduction to operational duties. "Sgt. Phibbs W. W." (as the operations order records) made his first operational flight 31 October, 1942. His pilot on that occasion was Sergeant Jimmy Lago who became his regular captain, and a colleague in the post-war era with the early Atlas Aviation company, and Spartan Air Services.

A little over two hours after takeoff on an anti-submarine sweep over the Bay of Biscay the crew spotted a red flare five miles from the aircraft. The Halifax was flying at one thousand feet. The crew could see five men in a yellow dinghy, one of them standing and waving. Trailing behind the dinghy was a green sea marker to enhance its visibility. Lago circled the dinghy but the crew kept losing sight of it in the low-hanging weather, once for a full hour. However, the patrol was able to alert rescue forces which arrived to pick up the men. After the crew of the Halifax watched the pickup they resumed their patrol, returning to base at 1842 hours. They had been airborne for 11 hours and 12 minutes.

If the novice flight engineer had any thoughts of exciting raids over enemy territory, they were quickly dispelled. The coastal flights, which normally lasted up to ten hours at a stretch, were reported tersely as "routine patrol". Such long stretches in the air offered one advantage to Sgt. Phipps, who could take over from his pilot and fly the bomber as much as he wanted without neglecting his primary duty monitoring the engine instruments. His actual time at the controls was not recorded but his eagerness to become a pilot would ensure that Jimmy Lago

## Whiskey Whiskey Papa

Freshly winged as a flight engineer and posted to 405 Squadron, Sergeant Phipps strikes a pose before an operational flight.

would be spelled off for hours at a time. Flying at low level demanded an alert man on the wheel to avoid the tendency after many hours in the air to fly into the ocean as some Coastal Command aircraft did.

Long hours of boredom did occasionally have flurries of excitement. "S Sugar" was patrolling at 3000 feet at 1521 hours on 16 November when the conning tower of a submarine was sighted in the Bay of Biscay. The U-boat was able to crash dive before the aircraft could get within striking range. After alerting another Halifax about the sub's location the patrol continued. That flight, captained by Sergeant Jimmy Burton, took off at 0823 hours and landed at 1905 hours, after 10 hours and 13 minutes aloft. Burton, an American from Billings, Montana, became the regular captain of Hugh Clee's crew. Weldy flew with Burton only twice, on 16 November and 16 December. He was teamed up for the rest of his tour with Lago. (Jimmy Lago was killed in 1953 at Dawson City, Yukon, while flying on aerial survey for Spartan Air Services. His boss at that time was his former flight engineer, Weldy Phipps.)

Patrols on the Bay of Biscay became very routine with nothing more than Spanish fishing boats reported for the many hours of scanning the ocean at low level.

On January 1, 1943, 405 Squadron was transferred from 4 Group back to 6 Group to resume bombing operations. Crews were taken off anti-submarine patrols and moved to Topcliffe. Within two weeks the unit moved again to Leeming, Yorkshire and then in April to Gransden Lodge, Bedfordshire. Its new assignment was as a Pathfinder unit, marking target areas to guide the main bomber force. It wasn't until the Phipps crew was ordered out on a shipping strike at Bordeaux, France, 21 January, 1943, that they got to see something of real action. Fog was reported and the bomb aimer had to calculate the target area. Bombs were dropped at 0333 hours from 18,000 feet. Explosions were seen on the east side of the Garonne River along the docks. Light flak was experienced around the target but the aircraft faced heavy scattered flak on the north side of the river. After return to base, the raid was declared a success.

> **A star is born...**
>
> If there was a lack of excitement in the war zone, there was related excitement one night back in Ottawa. Weldy's mother had gone to the movies. Before the main feature, a National Film Board newsreel showed crews of 405 Squadron being briefed before an operational flight. As the camera panned across one crew, the quiet theatre was jolted by a shout: "THAT'S WELLAND!" accompanied by an excited gesture to the screen that knocked the hat off the man in the row ahead. Mrs. Phipps had spotted her warrior son.

On February 1, the crew spotted an enemy aircraft with twin engines and a high tail only 1,000 feet away. As the hostile aircraft turned for the attack, Lago took evasive action into the cloud and lost the attacker.

The next event took place February 20 when they flew 10 hours and 14 minutes with the port undercarriage wheel in the down position. When the fuel had been burned down to safe levels all bombs were dumped into the ocean as a safety measure. The other wheel was lowered and locked successfully. The aircraft landed safely. The crew was to last through two more successful bombing runs, on March 11 and 22 on Stuttgart and St. Nazaire, respectively.

On 3 April, 1943, the Phipps Halifax took part in a five-bomber attack on Essen. It was Weldy's 27th operation. The record states: "Take-off commenced at 1953 hours. Three aircraft reached the target in good visibility and released their loads on PFF (target

## Whiskey Whiskey Papa

flares) from an average height of 17,000 feet. Actual bursts were observed by all crews. Tremendous blasts and fires were seen, the glow being visible for 120 miles on return. A particularly large explosion was observed by two crews at 2212 hours."

As the bomber worked its way back to England, anti-aircraft fire burst all round, finally scoring a direct hit that blasted away much of one wing. Lago fought the controls, holding full opposite aileron but losing his grip on the burning aircraft as it went into a shallow dive.

Recounting the event years later, Lago described Weldy as fearless. When bail-out became the only possible solution to the crisis, Weldy suggested that they hold on and ditch in the channel where they might be rescued. The urgency of his suggestion was based on an important date he had lined up back in England. Lago was not impressed. He ordered the crew to hit the silk while he could still maintain the bit of control he had. From his flight engineer's position, Weldy made sure all crewmen got out through the hatch. Then he escorted Jimmy Lago out and jumped into the night behind him. Clear of the stricken aircraft he pulled the ripcord on his chest pack. The chute deployed, part of the harness hitting him in the face and cracking a tooth.

The flight log recorded: "Two of our experienced crews are missing from this operation, being unreported from the time of take-off. Two of the specialist officers being Flight Lieutenant W. L. Murphy, gunnery leader, and Flight Lieutenant L.E. Luxford, bombing leader. Both are old timers with 405. Members of the missing crews (include) P/O Lago J. (pilot), F/O Colwell J.H. (navigator), Flt. Sgt. Beatty W.S. (bomb aimer), F/O Hoddinott W. H. (wireless operator / air gunner). Flt. Sgt. Waugh H. (air gunner), Sgt. Granbois A.B. (air gunner), Sgt Phipps W.W. (flight engineer). Operational flying time: 13:50 hrs."

*(It was learned later that one member of the Phipps crew was killed in the leap from the bomber. He was Flt. Sgt. Beatty of London, Ontario, whose body was never found.)*

Separated from his crew, Weldy parachuted to a safe landing. He hid his parachute then wandered carefully throughout the night knowing that the enemy would not be far away. Everywhere he went, however, he reached water. He changed direction, keeping his heading clearly in mind. But still he came to water. Before dawn he realized that he was on a small island. He had landed in Holland.

Shortly after dawn he met some civilians. He explained his eagerness to get back to England. But the locals were nervous with a fugitive in their midst. They explained that to get caught harbouring a flyer on the run would mean certain death for their families. Respecting their predicament, Weldy first considered his chances of evading capture without their help. He conceded that he could not possibly escape the heavy guard that was active in the area if he attempted to leave the island. He finally agreed to their suggestion that they turn him in to the German occupiers. At least they would get brownie points that might spare them harsh measures in the future. The German soldiers were right on the scene and he was taken prisoner. He was listed by the air force on casualty list No. 559 and became POW No. 1033.

In his book "Boys, Bombs and Brussels Sprouts" (Wing Commander) J. Douglas Harvey captured the essence of the downed flyer:

> It took a very dedicated and brave soul to suffer the destruction of his aircraft, parachute into the night, collect his wits, and then hide from the Germans. Alone, scared, unable to speak any of the languages, unsure of exactly where he was, or in which country for that matter, he had to hide by day and travel by night. Often he had lost his flying boots when he bailed out, or was not dressed for living outdoors in winter. Cold and hungry, he would call at the first farmhouse, hoping to get assistance. Often he was aided, but more often he was turned over to the German authorities, the fear of reprisal paramount in the farmer's mind.[1]

---

1 *Boys, Bombs and Brussels Sprouts*, by J. Douglas Harvey
 Used by permission, Mclelland & Stewart, Inc.. *The Canadian Publishers*

Once in the hands of the Germans, Weldy was delivered to a Dutch police station. As he and other crewmen anxiously awaited their fate, a German officer entered the room and, in perfect English, wished them a good morning. The prisoners were relieved to discover that their interrogator was a Luftwaffe officer instead of a member of the dreaded Gestapo. The officer removed his hat and placed it on the desk. According to Jimmy Lago, Weldy strode smartly across the room, grabbed the officer's hat and put it on his head, clicked his heels, gave the officer a Nazi salute and shouted "Heil Hitler!"

The officer quickly regained his composure and grinned, saying, "Gentlemen, I see you have a traitor in your midst." After that the questioning went well. But there is no record of ever meeting another of their captors with such a sense of humour.

~~~~~

**In the bag...**

The wartime map of German conquest overran international boundaries deep into the heartland of the Soviet Union. Central and Eastern Europe was dotted with prison and internment camps. Documents once classified secret show the map as if it had been sprinkled from a pepper shaker with centres for military prisoners of war and civilian internees. The military camps went by the titles of oflag, dulag, stalag and stalag luft, the latter being for air force POWs. An oflag was a facility for holding officers. The term comes from OFizier (officer) and LAGer for camp. A stalag for STAmmLAGer (stamm = main body) housed non-commissioned ranks. A dulag (DUchgangLAGer) was a transit facility. Near the end of the war, 185,631 British and 53,631 Americans were in custody. (All Commonwealth members were labelled British by the record keepers.) Survivors of wartime imprisonment refer to themselves by the germanic label "Kriegies" (for kriegsgefangener).

Among the many British captives was POW No. 1033, Sergeant W.W. Phipps. Following his capture Phipps was put on a train in Amsterdam and delivered to Stalag Luft III, located in a forest clearing near the town of Sagan, Poland. Stalag Luft III became the first of five prison camps he would endure during two years of captivity. The camp had a capacity of 5000 inmates but at war's end held 3240 British and 6750 American prisoners. It was from Stalag Luft III that the Great Escape of 76 prisoners took place in March 1944. The Gestapo executed 50 of the recaptured prisoners. Only three made good their escape.

The two buddies whose mutual support had been so solid were now separated for the first time since they joined 409 Squadron in February, 1942, as mechanics – but not for long. Three months after Weldy's Halifax was downed, Hugh's met the same fate. Both ended up at Sagan, a camp intended for officers only. NCOs could be accommodated if they served as batmen (servants) to those officers, an option that Sergeant Phipps flatly refused. "I was proud of him," Hugh said.

The rank-conscious Germans found a new home for Phipps and Clee. Stalag Luft I was located beside an airfield and large military barracks just northwest of the village of Barth-Vogelsang on the Baltic coast about 75 km northwest of the secret rocket research facility at Peenemunde. The immediate area included a training camp for German troops, dampening the prospects of an easy escape. But escape was a hope active in most prisoners' minds. In every camp, tunneling was a regular feature. Weldy took part in tunneling at one or more camps and worked as a "miner" digging underground and a "penguin",

### Achtung! Schpitfeuer!!

About the last thing the Germans wanted to see in a prison camp was a Spitfire. But that's what Weldy needed to get him into the "flying club" at one camp in which he was held. A veteran model builder, he was not about to be excluded from such recreation. In a letter home he asked that his cousin John Deacon send him a model. "I sent it through the Red Cross," John said, "and it got there." There was no word of its effect on Weldy's captors.

carrying excavated sand outdoors in his pants and releasing it as he walked around the compound "gardening". In later years Jimmy Lago described him as a great tunneler.

~~~~~~~

In November 1943, it was time to move again. This time 1000 prisoners took a 1000 km. train ride to the northeast. Travel arranged by the German guards was far from luxurious: accommodation was in ancient boxcars left over from the First World War. Each was inscribed with its capacity: forty men or eight horses. Normal loading when prisoners were being moved was about 80 per car. There was little or no food, water or warmth. Sanitary facilities consisted of a drum that quickly filled and slopped as the train lurched along the tracks. The only ventilation came through two barred windows 12 by 18 inches in size. There was no room for all passengers to sit, leaving most of the men to stand much of the way. The new destination was Stalag Luft VI at Hydekrug near Konigsberg, East Prussia. Again, the German map claimed the area, which was actually just on the Lithuanian border.

Escape tunnels were constantly under construction. The hope was to get clear of the camp and find some of the Polish workers, enslaved by the Germans but living outside any prison gates. Trustworthy operatives might get a fugitive onto a boat for Sweden. That idea gained popularity when it was learned that Hitler had decided to execute all prisoners. That plan was believed to have been rejected by Goering who feared similar treatment of German prisoners in retribution.

At Hydekrug the two buddies found each other again. Weldy had been lodged in Lager J while Hugh was in Lager K. When Weldy's whereabouts became known, Hambone put in a request to camp authorities to transfer his friend Cub into the same lager. The move was approved. In recent years, Hugh laughed about his role as father figure to Weldy. "I was 20 and he was 19 when we first met, hardly a big enough age difference for fatherly guidance."

In Lager K the duo formed a 'combine', a pairing that shared food, companionship and dreams of escape. Hambone and Cub had such respect for each other that if all else failed they could operate as a two-man team against all the privations the Germans could throw at them. It was a brave assumption early in their imprisonment but one borne out in the later events of their ordeal. Phipps was described by fellow prisoner Ed Houston as a man totally without fear. Hugh Clee might have modified the description slightly. "We used to walk around the compound looking at the warning wire just inside the fence," Hugh said. "I'd say to Cub 'let's jump the wire and make a run for it.' The trouble was, I'd just be kidding but he wouldn't think twice about it. It was well understood that crossing the warning line meant certain death." The warning wire was about one metre high and set inward of two outer fences, 10 and 12 feet high respectively.

As much as the various camps imposed hardship and general discomfort on the inmates, there have always been memories of the humour generated in the confines of the prison. The diet was never one to tempt an appetite. Since cabbage was a staple food, there was the frequent cloud of methane gas in the hut, much to the displeasure of those walking into the flatulencia that drifted around the rooms. But such clouds had a silver lining of sorts. On one occasion a newcomer to a camp was fascinated by the howls of laughter coming in bursts from the next room. On investigation he found that the prisoners had found a very satisfying form of entertainment in 'blue blazers'. In succession, the men would issue a cabbage-generated fart while another held a match nearby. The result would be a burst of flame and a wild burst of laughter in appreciation of the feat.

There were other negative effects from the food. Some combinations in the sparse diet brought about the frequent need for urination, especially after lights out. This created a hardship for the men sleeping in three-tiered bunks: it meant climbing down from the compact bunks and finding the "bog". By any standard it would be hard to recognize the half barrels as toilets. The inconvenience was too much for one of the bunkmates. Carl "Soggy" Norton, a former Ottawa policeman and Rough Rider football player. He devised a pipeline that ran from the top bunk, where his oversized frame rested, down to a private pail at the floor. This tube, constructed from scrap Klim (powdered milk) tins worked well for him but brought complaints from the lower levels. One wonders how this top bunk innovator earned the name Soggy when the complaints came from the bottom bunk.

Toilet facilities during the day were tolerable but the only night facility was a single bucket in each sleeping room. Since many men were afflicted with diarrhea, the indoor latrine was totally inadequate. Hygiene could hardly be maintained with one shower every two weeks – not enough to discourage the usual infestation of lice, fleas, ticks and bedbugs let alone control body odour.

In its final days, the Hydekrug camp, which had been designed to accommodate 7000 prisoners, actually contained 4717 British (including Canadians) and 2411 Americans. The density was much lower than in other prisons but if that offered any comfort at all it was soon to be overtaken by what was to come.

News of the Allied war effort was hard to come by, but brief news reports from the BBC kept hopes of survival alive. The news was received by a clandestine radio the prisoners had built from a collection of parts. The Germans seemed to know that it existed, resulting in frequent searches of the huts. It had been hidden on such occasions, however, so that it was never found. The prisoners would relax and smile at the guards as they ransacked the huts, looking in every possible hiding place. Where was it hidden? That was a question that Weldy would not answer even years after the war. It became a recurring question by Bob Fowler, a fellow pilot in the postwar years. Weldy teased Bob by refusing to tell him its exact location. He never found out.

Shortly after D-Day the Russian armies had advanced south-westward, injecting a serious element of anxiety into the hitherto confident Germans. The threat finally triggered an order to move all prisoners inland ahead of the Red Army. It was a measure to guard against losing the bargaining chip the POWs represented.

There had been camp scuttlebutt that prisoners would be moved to prime target areas such as Berlin in order to discourage bombing raids. The Germans were not anxious to lose such an opportunity. Accordingly, prisoners at Hydekrug were marched to the nearby rail yards where they were herded once again into the creaky man/horse boxcars.

The cars were sealed shut in the stifling heat; only the small barred windows admitted any air. The crowded conditions made any form of comfort impossible. The men learned to sit with knees up to provide backrests for their fellow travelers. Sitting had to be done in shifts. The only food was what each might have carried from the camp. The lack of sanitary facilities resulted in a horrible mess particularly since many of the prisoners had dysentery. The trip inched along for three days and three nights.

The foul conditions aboard the train intensified the desire to escape. At least three prisoners managed to saw bars off a window with hidden hacksaw blades and gain temporary freedom. Others failed in an effort to saw through the floor of a boxcar. To cover the sound of sawing, the prisoners formed a chorus, singing loudly. The car carrying Weldy Phipps, Hugh Clee, Ed Houston and Soggy Norton was guarded by a rather nonchalant young soldier who caught the spirit of "You Are My Sunshine". He pulled out a harmonica and provided accompaniment. While the guard was preoccupied with the music, Soggy removed the magazine from his rifle and dropped all the bullets down through a hole in the boxcar floor. When the guard discovered the ruse he burst into tears, pleading that his superiors not be told. The softest punishment he could get was a posting to the dreaded Russian front. Camp life had seen several guards compromised, usually with chocolate or cigarettes from Red Cross parcels. A compromised guard provided a good source of escape materials or relaxed surveillance.

When the train arrived at the seaport of Memel on the Baltic the prisoners were marched in stifling heat to a boat called the Insterberg, a small coal carrier built in 1919. The men descended into the pitch black hold on two slender swinging ladders, stirring a cloud of coal dust as they moved. Unable to see, the prisoners stumbled all over their comrades. When about 500 men had been taken aboard they shouted to the guards that there was no more room. But hundreds more were forced into the boat. Seating was impossible in the crowded belly of the vessel. Added to the physical torment was the thought of friendly fire: an Allied submarine might send a torpedo into the hull or the vessel might hit a mine. Air force colleagues might find it to be an easy sitting duck.

The voyage took three days and three nights. Food, water and sanitary facilities were not high on the Germans' list of cruise ship amenities. Water was lowered in a bucket through two tiny hatches some 30 feet above the heads of the passengers. But the rocking motion of the high-riding vessel made the bucket swing, slopping much of its contents and denying many a badly needed drink. The bucket was returned filled to the brim with urine which otherwise just went into the bilge with the ejecta of seasickness. Those contents, too, slopped onto the passengers on the way back to the deck. The boat finally docked at Schweinemunde, occupied Poland (today: Swinoujscie), 150 km north-northeast of Berlin.

The arrival was greeted with heavy shellfire, directed at a flight of American bombers, from the battleships Liepzig and Prinz Eugen. The barrage was supplemented by the guns on the Tirpitz, which had been sunk but rested upright on the harbour floor with its upper decks still above water. The prisoners were assembled in a rail yard and feared for their lives as they assumed the railway complex was the intended target of the American bombs. The Germans blanketed the area with an acrid smokescreen. Fortunately, the bombs fell away from the prisoners.

## The Run UP The Road...

The departure from Schweinemunde 19 July, 1944, marked a significant worsening of the POW experience for the former Hydekrug prisoners. Once again they boarded a horse train for the 200-kilometre trip to Stalag Luft IV at Gross Tychow, Poland. "When we arrived at the station in the morning, we were made to remain in the boxcars until noon, holding our kit," Weldy explained following his return to England. "We were chained in pairs or in three and fours still holding our kit when we were allowed out of the train."

They were now forced to march the final two miles to their new destination. And for this phase of the move, the Kriegsmarina, young marines of the Hitler Youth element, took over guard duties. They were led by a maniacal Hauptman (Captain) Pickhardt, who screamed hysterically at his troops and at the prisoners. Apart from his brutal manner he could hardly be seen as the quintessential Nazi. He wore white breeches and gleaming black jackboots. His face was heavily plastered with cosmetics and he smelled as if doused in perfume. His favourite weapons were bayonets and Alsatian guard dogs.

From the Phipps report: "A large number of guards were assigned to us and they lined the two-mile road which led through a small forest to our camp. Guards with fixed bayonets marched alongside us with a squad of them front and rear. Hauptman Pickhardt gave orders for the guards to use their bayonets, then started us running. It was impossible to carry our kits and we dropped them on the way. The guards jabbed us constantly with bayonets as we ran and stumbled along. They let dogs loose among us and many of us were bitten."

The young marines entertained themselves by prodding the prisoners and then comparing the depth of blood marks on their bayonets. One Canadian was carrying a banjo when a guard set a dog onto him. He smashed the banjo over the head of the animal sending it into a wild frenzy. It turned on its guard handler and ripped him savagely.

"Some chaps were chained in such a way that they were unable to drop their bags and they stumbled and were bayoneted frequently," Weldy said. Hugh Clee confirmed that in addition to using bayonets as torture, the guards would try to slash the packs from the backs of the prisoners in order to steal their possessions. One of the attacks from behind came in the form of a rifle butt over Hugh Clee's head, splitting his scalp and buckling his knees. Six-foot-three-inch Soggy Norton, who had been chained to a very short Polish prisoner, was having difficulty moving fast enough. Then a guard bludgeoned the Pole, knocking him out. Soggy put the two packs under one arm and the unconscious Pole under the other and ran. Both Weldy Phipps and Hugh Clee were stabbed during the hour-long run. Both lost their kits.

"From the rear the Huns fired shots over our heads to create a panic," Weldy said, "to get us to break ranks so they could shoot us." In the adjacent bushes, the prisoners could see machine guns mounted with cameras ready to photograph any breaking of ranks. Film showing the slaughter of escaping prisoners could provide proof that lethal force was justified. The prisoners would not break ranks although the main topic of conversation at all times was escape. "When we arrived at the camp we had nothing with us but the clothing we had on. They handled us roughly when they searched us and were particularly tough with several Jewish fellows among us." The guards searched for radio parts or weapons that might have been dismantled and distributed among prisoners. Even individual buttons were examined.

Entry to the camp at Gross Tychow was through a narrow passageway. The Germans put guards on each side of the path, and forced the prisoners to run the gauntlet, jabbing at them with bayonets as they struggled through. Prisoners able to keep up and move fast enough caused a confusion of closely spaced rifles that spared many of them further injury. None of the prisoners had eaten or slept for three days. That particular forced march became known by surviving prisoners as The Run Up The Road.

"Pickhardt was a madman," Weldy said. "One day he tested all the machine guns on the walls, firing at targets directly over our heads, probably to impress us with the effectiveness of the guns. Camp rules were rigidly enforced and men were shot at for the slightest infraction." (It was not until well after the war that Pickhardt's fate was known: the Russian army overran his position and he was shot dead in his tracks.)

In February 1945, the Red Army moved uncomfortably close once again. Some 4000 prisoners were formed up for a march from Stalag Luft IV toward the southwest. As the column proceeded, the land war intensified. Flashes of artillery fire at night generated contrasting emotions in the prisoners and the older veteran guards that had replaced the rabid Kriegsmarina. The Russians were advancing from the east and Allied forces were squeezing down from the north. The column had moved some 750 km when they halted temporarily 75 km south of Hamburg at Stalag 11B, Fallingbostel. The stop was brief, about 10 days. For reasons unexplained, the march was restarted but directed east, back towards the Russians. Along the way the prisoners saw evidence of fresh destruction inflicted by Allied air forces: burnt-out vehicles and a train blasted by rocket fire from Typhoons.

At this point in the march some prisoners reported a better supply of food. One marcher felt privileged to have salvaged a horse's lung, which he grilled. On April 15th, the column was routed out for a 2:30 a.m. resumption of the march. The guards' anxiety and the growing confusion offered a better opportunity for a break from the column. Three days after the turn back east, Weldy and Hugh broke ranks and dashed into the nearby forest. There they built a lean-to with fir branches, and lined the floor with boughs for bedding. This accommodation blended into the background and went unnoticed by the German soldiers still active in the area.

While the bivouac provided a place for a badly needed rest from walking, running and stumbling along the route, there still was the problem of nourishment. "Food was the grimmest part of the march," Weldy reported. "We had virtually no rations and lived off whatever we could pick up. I lost between 25 and 30 pounds in weight." Hugh Clee, who had studied German in high school, was able to scrounge or barter for food along the way. When they escaped from the column their larder was down to a sock full of dried peas, a sock full of flour and one third of a loaf of black bread.

When they thought it was safe to take a look around for guards, Hugh sneaked out to the edge of the forest. As he lay on the ground he could see some guards having lunch. A sergeant walked right towards him, stopped and urinated not more than 15 feet away. "I couldn't believe that he didn't see me," Hugh said, "but he didn't. He could have heard my heart beating."

After five or six days in the woods, all hell suddenly broke loose. Heavy gunfire sent shells ripping through the trees from both directions. Soon the fugitives could see more vehicular activity: some horse-drawn wagons and some tanks. The tanks posed something of a question because they bore large white stars on their sides. The airmen were unfamiliar with armoured markings – unaware that the white stars meant Allies.

Deciding that the tanks were not German, the pair approached the roadway cautiously. Immediately they were looking once more at the business end of guns. The tank crews had no quick means of identifying them and appeared bent on shooting them rather than entering into formal negotiations. Quickly the fugitives took off their undershirts and waved the "white flags" at the menacing tankmen. Although the shirts were far from their original white, the universal surrender signal worked. "We were a rag-tag pair," Hugh Clee recalled. "We looked like the Polish slave labourers we had seen along the way."

The two escapees were taken aboard the tanks – freedom at last! But not quite: once aboard the tanks they roared into battle. Then the tanks were formed into a "hedgehog", an outward-facing circle of the wagons for protection. The newly liberated airmen were in good hands – the 7th Armoured Division of the 2nd British Army, the "Desert Rats". For the two flyers the war was finally over. For Weldy it was the end of two years and 12 days of captivity.

"We asked the British for grub," Weldy said. During lulls in the action, the crewmen fed their passengers with bully beef and canned plum duff. The beef was pulled from the cans and devoured much like one might eat an apple. Stomachs unfamiliar with that much food, and particularly such rich food, did not react well, especially after such an exciting day.

The tankmen delivered their passengers to Celle, 50 kilometres to the southeast, for a thorough delousing. It took more than one treatment, including a head shave, to rid them of their ticks, fleas and lice. From Celle they were transported to Brussels for the short flight back to England. Weldy Phipps, who had departed from England as a sergeant returned as a Warrant Officer Class 1, 24 April, 1945, having gained three promotions in his two years of incarceration.

During the late part of the war, authorities had scripted a policy to deal with LPWs – liberated prisoners of war. There was to be no more combat for them and every effort was to be made for their early return to Canada. Top priority for repatriation went to LPWs with more than six months' imprisonment or at least three years' service

Warrant Officer Phipps, just home from two years as a POW, had an unusual experience when he and sister Gerry tried to locate old friends.

overseas. Much was made of the psychological condition of the returned POWs, their likely impatience to get home. Weldy was put aboard an aircraft and flown back to Canada, landing at Dorval six days after VE Day.

## Sprechen sie Kriegie?

After his return to Canada Weldy and his sister Gerry decided to visit old friends in the Ottawa Valley. It wasn't as easy a navigation job as he thought. After criss-crossing several rural concession roads in the Richmond area they failed to find their friends' farm. Weldy spotted some farm workers and approached them to ask directions. But when he got closer he saw that each was wearing coveralls with a big red circle on the back. He had no trouble recognizing German prisoners of war. And they had no trouble recognizing his Warrant Officer's uniform. Somewhat intimidated, they shied away from his inquiries. But when Weldy spoke to them in German he had picked up in his two years as a POW, the directions came willingly. Still suffering the discomfort of bayonet wounds and dog bites he got while a prisoner of the Germans, he expressed no resentment at the relatively soft treatment of the German prisoners.

~~~~~~

## Chapter 2
### Picking up the pieces...

With the war in Europe behind him, Weldy signed on with the interim air force that was to continue the fight in the Far East. That was a standby arrangement that would only engage his services on an if-and-when basis. The atomic bomb removed any such necessity and so Warrant Officer Phipps set out with his new rank as Mister, his thoughts focused clearly on the future.

His love affair with airplanes had not cooled in the least. In his discharge documentation he signified his intention to study aeronautical engineering. He had worked hard to prepare himself for such advanced education while in prison camps, earning eight senior matriculation credits. The curriculum was one issued by the University of Toronto and delivered through the Canadian Legion Educational Services. To complete his matric he enrolled at Ottawa's Nepean High School in the fall of 1945, renewing his study of mathematics. But it did not come easily. His math teacher, Mr. A.T. Appleton, recognizing his determination, tutored him at home until he succeeded in his exams.

Civilian life as an adult student had other problems. School regulations forbade smoking within a block of the premises. And Weldy, along with half the class of veterans, was a heavy smoker. But there was one diversion from the cut into his smoking time. It was the fuselage of a Tiger Moth in the basement of the school. Bill Law recalls his classmate sitting in the cockpit, talking to himself as he 'flew' the aircraft. That play-acting helped deflect his career from engineering to hands-on flying.

Armed with a bundle of back pay from his two years in prison camp, Weldy had presented himself to the Ottawa Flying Club January 2, 1946, to take his first official flying lesson. He climbed into a war-surplus Tiger Moth with an instructor and took off for the first 30 minutes of a flying career that ultimately filled more than 18,000 airborne hours. His first solo flight was just eight days later, after 10 hours and 55 minutes of instruction. Given his motivation, one can only imagine the impatience with which the student pilot endured the wait to be set free.

Now it was time to step up. On March 3rd he applied his Tiger Moth experience to the Piper J-3 Cub. After five minutes' air instruction he left instructor Hugh McGuire on the ground and flew off with the Cub on his own. His next milestone was a licence and 25 solo hours, which would permit him to carry passengers. Somehow he overshot the mark by 40 minutes but he was able to give a classmate, Joe Holmes, a 50-minute flight May 19th.

"He said he needed ballast in aircraft and so he put me in the front seat," Joe said. "We climbed up to 5000 feet. Weldy then put the Cub into a wild spin directly above the town of Manotick. Weldy explained that he had to practice spins." He did not explain that such practice with passengers was illegal. On the same day he took his brother Gordon for two flips lasting about an hour in total. The next day he gave Bill Law a ride. That also turned out to be thrilling. Weldy demonstrated another spin. Bill was impressed. The experience convinced him to take up flying so that he, too, could some day terrify his friends.

Atlas Aviation, the pathfinder company Weldy was to establish out of Resolute Bay in 1962, had two lives. The second was not exactly a clone of the first but there was a sentimental connection between them. It was Atlas No. 1 that gave Weldy his first job in civil aviation. He was hired in 1946 as aero mechanic, a term later upgraded to aircraft maintenance engineer.

President of the original Atlas was Angus Morrison, a veteran of tank battles in North Africa and Italy, a war that diverted his mind towards some form of transportation less earth-bound-clumsy than armoured vehicles. His interests turned skyward. Aircraft fascinated him. Artillery units used spotter aircraft but the tank units did not. Why not? And so Captain Morrison put the question convincingly to his commanding officer. The CO authorized a flight in an Auster aircraft used by the artillerymen. Angus was smitten.

Before the war ended, Captain Morrison had to be evacuated to England by hospital ship. When his health permitted he was returned to Canada where the army found employment for him as a staff officer in the Adjutant General's office. It was a job he hated; his main interest was still flying and he was determined to fly.

As soon as the war was over and civilian life began to resume it peacetime functions, Angus reported to the Ottawa Flying Club for his first lesson. "I was in the front seat of a war-surplus Tiger Moth," he recalls. "The instructor pushed on the power and we started down runway 22. Soon we were at 200 feet and headed for a crossing of the Rideau River. Then the engine quit. I was staring at a propeller that was dead still, nothing but the whoosh of air over the canopy. I looked at the steep banks of the river and braced for the crash. The instructor shouted: 'Just relax!'" Angus did not manage to relax but the aircraft did get manoeuvred back onto the ground in one piece. He cancelled the rest of the lessons.

The Ottawa airfield also accommodated the McGuire Aviation School. McGuire was equipped with two brand new Piper J-3 Cubs. Such aircraft, unabused by the sprog pilots of the wartime air force, seemed like a safer bet. And so Angus resumed flying lessons under the coaching of Hugh McGuire, with whom a compatible student-instructor relationship developed. He soloed in April 1946, near the time he was to be discharged from the army. That brought up the inevitable question: "What are you going to do?" Angus didn't know. But he did allow that he wanted to fly. And so Hugh suggested that he invest in the McGuire Aviation School and become a partner. It was not a decision Angus had to ponder for long: he had inherited some money, and as of August he was finally out of the active army. He became a part of the re-emerging civil aviation scene.

The McGuire aircraft inventory was increased by one more Cub and a sleek all-metal Globe Swift. The partnership was increased by the addition of Scott McLean as a flying instructor. It was, however, a short-lived collaboration. Cancer claimed

Hugh McGuire only a couple of months later. The remaining partners renamed the company Atlas Aviation Limited and set up operations in the administrative building of Laurentian Air Services. Aircraft were housed in the former RCAF No. 3 Hangar which was shared with Spartan Air Services.

With the addition of war-surplus and new aircraft, Atlas needed a skilled maintenance engineer to take care of them. And that's where Weldy Phipps came in. He had all the qualifications for the job. But mechanical work was only half way towards the dream he had nurtured from boyhood: he wanted to get to the top of the aviation ladder. The 251 flying hours he spent on wartime operations allowed him lots of time relieving the pilot of his Halifax bomber. But much of that was unofficial pilot time. Atlas appealed to him as the ideal doorway to becoming a pilot. He was offered the job at $60 a month and all the flying time he could possibly log.

Weldy was a natural who required very little drilling to refine the unusual coordination that was to sustain him eventually through 28 years as a pilot's pilot. "He took to flying like a duck to water," Angus Morrison said.

At the end of six months as a private pilot Weldy had accumulated a total of 59 hours and 55 minutes, of which 12 hours 45 minutes was dual instruction. He was checked out on the Globe Swift July 26 after several short dual flights with Scott McLean. During the summer of 1946 his log book shows numerous "test flights" with Atlas Aviation, an indication of his "pay" for technical services. August 19th, 1946 was not officially Mother's Day but it was in Weldy's new life: that was the day he took his mother flying in the Globe Swift.

About a month later a young lady appeared at the fence to watch the aircraft come and go. The striking brunette drew the attention of a young man who inspected her from the hangar office window. Was she a prospective student? He wasted no time making her acquaintance. Approaching her, he asked: "Can you type?" She said she could and so was invited inside the fence and into the office of Atlas Aviation. Sitting at the typewriter, Frances Coolin pecked on the keys as the handsome young man dictated the letter. She was nervous and made several mistakes, including the spelling of 'hanger' when it should have been 'hangar'.

"What's your name?" she asked the author of the letter.

"Joe Smith," he said.

"Sounds phony to me," she said. And when pressed, he admitted to his real name.

Frances had borrowed her sister's bike to go to the airport and so Weldy, flushed with the success of his opening gambit, volunteered to drive her home in a car she later described as an "old jalopy that had its fenders held on with haywire". The car was a 1928 Huppmobile acquired from a member of the Flying Club just for getting it running and taking it away. He haywired the bike to the back of the car for the trip back to town. On September 26th, Weldy's log book records a 30-minute flight in a J-3 Cub. His passenger was "Franny". On October 5th she got two rides. Three weeks later she got a ride in the Globe Swift.

Since both were airplane nuts, Weldy undertook to teach Franny to fly. But she was too much in awe of her instructor to concentrate on the instructions. On the other hand, he felt intimidated when he issued the firm orders an instructor often has to give to a student pilot. Weldy proposed a deal: "Let's get married and I'll do all the flying." Fran gave up her piloting ideas and they were married March 29, 1947, six months after they met at the airport fence. Weldy's small pay cheque posed a

Wedding bells rang out for Weldy and Fran six months after they met at the airport fence.
--Courtesy Geraldine Phipps Trudel

## Whiskey Whiskey Papa

big problem: he was determined to press on with more training and that required some funding. Fran made $90 a month at Ottawa Hydro and that put a dent in Weldy's pride. Atlas celebrated the marriage by raising his pay to $100 a month.

~~~~~~

### Storming the barns…

With increased domestic responsibilities looming, Weldy expanded his flying agenda. He wanted to build his log book with flying hours and pocket some extra cash at the same time. He had earned his private pilot's licence May 16, 1946, and his aircraft maintenance engineer's ticket on June 26 that year. His commercial licence followed in 1947. Now he could do some legal barnstorming. Much of this activity took place on the weekends – landing in rural fields, attracting spectators and then offering quick circuits to paying customers. The locations were always within an hour's flying time of Ottawa, including the Ontario communities of Smiths Falls, Fitzroy Harbour, and Luskville, Bouchette, Gracefield, Shawville and Maniwaki in Quebec.

These operations were undertaken on a team basis. Weldy would do the flying while his buddy, Bill Law, went along as parachute jumper and hustler. He also conducted passengers from the rear of the plane to the seat, making sure nobody got hit by the turning propeller. Often Lorna Bray (later celebrated pilot Lorna deBlicquy) went along in her own effort to gain flying experience. She learned at an early age that when it came to flying, men came first and she often only got to drive the car.

A normal barnstorming operation would begin with a flight at about 300 feet over a community to attract attention. Often the spectacular descent of a parachutist would then establish where the aircraft would land. The crowd was quick to assemble. Then the pair went into action like snake oil salesmen. For three dollars, a customer could take a six or seven minute flight, a bit more if they wanted to see their own farm or house. Weldy made sure the aircraft flew out of sight of the crowd to give the impression that the flight was longer than it really was. That increased the urge to fly in first flighters. The line-up often had at least 30 passengers waiting, money in hand.

While the weekend flights brought in money, they were not without hazards and various forms of excitement. Several thrills came from the choice of unprepared fields from which they operated. Often Weldy would perform a steep sideslip into a field, then the two would get out and clear debris from the area. It was his custom to offer the farmer five dollars for the privilege of using the land. It was a bit more when a hay crop had to be cut to let the aircraft operate.

~~~~~~

## Oh no, not again...

One of the most memorable landings in the name of barnstorming came in a farmer's field north of Ottawa. It looked clear of obstacles and livestock, and so Weldy circled and landed, looking forward to a group of eager townsfolk arriving for a ride. But this farmer had other ideas. He confronted the Cub's crew, shotgun in hand. "Get the hell offa my property," he shouted. The barnstormers looked anxiously at the gun. Weldy, not wanting to have a second aircraft shot out from under him, agreed to depart in haste. Some time later it was discovered that the irate farmer was related to Fran.

~~~~~~

## An aerial butt kick...

The volatility of aviation gas is one of the first things a new pilot learns. And nobody knew it better than Bill Law who was a student of engineering during much of the barnstorming era. Riding in the back seat of a J-3 Cub, he would carry cans of avgas on his knee. The record was five five-gallon cans, but this proved extremely dangerous in an aircraft that had a baggage capacity of 24 pounds. A take-off run of almost 3000 feet and a negligible climb, resulted in subsequent gas loads being reduced to three five-gallon cans.

Adding to the danger was Weldy's serious smoking habit. Bill nearly had a heart attack on one flight when the smell of smoke and leaking avgas filled the small tandem cockpit. Weldy got a clout on the back of his head. He tossed the cigarette out the window.

~~~~~~

### Escape and evasion...

On another occasion Bill had three cans on his lap when the pair decided to check out the possibility of hopping passengers out of Pendleton airport, a retired wartime field about 30 miles east of Ottawa. Pendleton offered a triangle of tarmac runways much safer than rough fields. As soon as the Cub touched down, a car raced up, its driver threatening arrest for landing at a closed military field. There is an aviation equivalent to the marine any-old-port-in-a-storm provision and Weldy was quick to exercise it.

"It's an emergency," he declared. "We're out of gas." Bill took the cue and struggled out of the back seat with a five-gallon can. The Cub had not burned much gas in the brief flight from Ottawa. Bill, nonetheless, began to fill the "empty" tank which was situated right above the engine. He still shudders as he recalls the gas sizzling on the hot exhaust manifold. The guard, obviously not a flyer, accepted the "emergency" as legitimate and let the Cub depart.

~~~~~~

### Unwelcome welcome home...

Ground damage to the aircraft was not uncommon. On one landing near Fitzroy Harbour, the Cub's prop hit a cable that had been left in a field. The evening sky was darkening as the Cub fought a 40 mph headwind, reducing headway to slightly better than a standstill on the way home. When they finally reached Uplands airport Weldy made a straight-in approach. The tower was not advised because the Cub had no radio. But Weldy was unconcerned, since Uplands airport had little traffic and no airline arrivals at that time of day. He taxied to the Atlas hangar directly across from the tower and phoned to announce the arrival. The controller wanted to know how he got into the airport unseen. In addition to having no radio, the Cub had no navigation lights. Weldy later received a violation from the Department of Transport for landing after dark. The damaged prop had to be scrapped.

~~~~~~

## The shear nerve...

Weldy's sister Gerry was a silent partner in many of his enterprises. But there was one time when her silence was broken. It happened in the early fifties when her brother wanted to refabric the wing of his personal Fleet Finch. Gerry was then a partner in the Ottawa Feather and Mattress Company. What better place than the work tables of a mattress company to reskin a wing?

Weldy carried the skeletal wing into the plant, laid it out and proceeded to prepare the fabric for it. For the cutting he borrowed Gerry's prized pinking shears, a birthday gift from her father. The fabric was cut and put on the frame with 'banana oil' dope. The result was a skin tight wing and a gummed up set of pinking shears. "I don't even know what became of the shears after that!" she said.

~~~~~~

## Here's lookin' at ya Wakefield...

Cables, cattle, fences and tree stumps were not the only hazards to successful barnstorming. Weather could play dangerous tricks on barnstormers who extended their Sunday afternoons one trip too many. Weldy and Bill were returning to Ottawa from Bouchette, Quebec, late one afternoon. Bill, as usual, was nursing three empty gas cans. To gain some comfort, he had left the seat belt undone. "Suddenly, we hit a thunderstorm and just dropped out of the sky," he recalls. "I was pinned against the roof with the gas cans.

"Weldy regained control and by pure luck we broke out of the overcast 100 feet above the Gatineau River. We followed the river staying just below the clouds. As we passed the town of Wakefield we could see people at eye level looking at us in disbelief." The incident gave Weldy a scare that he filed away as a safety lesson. Bill was so busy trying to find his seat belt that his scare didn't take effect until after they had landed.

~~~~~~

## Mass of a different kind...

The barnstormers could not really claim to have airlifted celebrities during their freewheeling operations but, on a rural level, they did come close. North of Ottawa the locals had taken up a collection so their parish priest could have a flight. It seemed like a good idea at the time, but when Weldy got the passenger into the aircraft he feared for the little Cub's ability to lift the load off the ground. Bill tried to close the door on the plane. He had a lot of trouble. "The priest must have weighed 250 pounds. He just oozed over the edge of the Cub."

Not one to give up easily, Weldy decided he might have a better chance of taking off if he made a downhill run. But the Cub refused to leave the ground. Embarrassed, Weldy taxied back and refunded the money.

~~~~~~

## Jump start (and finish)...

Weldy's romance with the parachute was a matter of survival, excitement, and business promotion. When he got into civil aviation, he again wanted to try parachuting. He bought a war surplus parachute from Flight Sergeant George Bennett, a veteran of 1300 jumps. George taught Weldy how to pack a 'chute.

On July 1st, 1947, he made his first demonstration jump in friendly skies at the Carp Air Show. The Ottawa Journal reported that he jumped from 2000 feet and landed one minute and 21

Weldy's non-combat parachuting was short-lived after his primary 'chute failed in a jump at the Carp air show.
--Phipps collection

seconds later. "The crowd cheered as the orange striped parachute slowly mushroomed out allowing Phipps to float slowly to earth." Weldy reported: "It was beautiful."

The success of his stunt led to other jumps. On July 20th he was barnstorming at Bouchette, Quebec. To attract a crowd he jumped from a Piper Cub from 2000 feet. He floated down, narrowly missing some high tension power lines and landed right on top of a bird's nest. Public fascination with the display paid off. He logged five hours and 35 minutes selling rides to the gallery that had gathered.

The jumpers wore nothing in the way of safety equipment – no helmets, no boots; just coveralls over street clothes. The only precaution was to remove everything from the pockets so nothing would be lost in landing.

On July 24th and 25th Weldy made practice jumps from 1300 and 900 feet at the Ottawa Airport. He had requested and got permission to jump into the middle of the airfield. He didn't want to land on pavement and was lucky enough to miss it by three feet. A line of cars drove out onto the field to greet him and retrieve the 'chute.

On July 28th he made two jumps at Carp Airport. The first one was from 2000 feet. He repacked the parachute and offered it to Russ Bradley for the next descent. Then he changed his mind. He hadn't liked the twist in one of the risers and decided to make the jump himself. He boarded T.V. Little's Piper Super Cruiser. The plane climbed to 3000 feet. Weldy posed at the door to bail out. After a free fall that wowed the crowd below, he pulled the ripcord. Nothing happened. He had five or six long seconds to contemplate his mortality. He grabbed the D-ring on the reserve pack and gave it a powerful yank. The second chute snapped open. The crowd, including Fran who was busy snapping pictures of the death defying leap, knew nothing of the chute failure. Summarizing his flying career later, Weldy confessed that he found something more sensible to do than jumping out of airplanes. He never jumped again.

~~~~~~

## Pogo the whirlybird...

After his experience with Beaufighters and Halifax bombers, maintaining Piper Cubs, Ansons and later Beaver aircraft was a breeze for Weldy. There was one challenge, however, that depended more on his Ottawa Valley haywire style of maintenance: the KD1A autogyro Atlas had acquired on a whim. While reading an advertising sheet called Trade-a-Plane, Angus and Scott spotted this unusual machine for sale in Kalamazoo, Michigan. It struck the business-hungry entrepreneurs as a good way to make money – towing banners, if nothing else. And so the KD1A was purchased and moved to Ottawa. It was a much speedier transfer than flight for the ungainly aircraft: the purchasers merely folded its rotors, loaded it into the back of a truck and hit the highway. It wasn't the first exciting trip for the machine. It had served its early years on the Texas Border Patrol spotting migrant workers trying to sneak into the United States from Mexico.

    The autogyro presented a challenge to any pilot or mechanic. A little research and some luck turned up Kellett, the KD1A designer-builder. His advice for the prospective pilot included a caution that the flimsy aircraft had never had a stress analysis. To fly it safely the pilot, seated in the open cockpit, was advised to position the aircraft so that the wind blew over his left shoulder. Then the engine was started and the overhead rotor powered to 200 rpm by the engagement of a Model A Ford clutch. The Jacobs engine was revved to 2000 RPM. When those readings were achieved, the aircraft was to be turned into wind. It would hop into the air.

"Pogo" required some unusual maintenance. Frequently Weldy would have to drag it into the hangar to have its rotor blades tuned. This was done by checking the balance of one of the three blades against the other two. To bring them into equal balance lead shot was poured into the fabric-covered blades through a small hole near the tip. When the balance was achieved, the rotor would not shake the frail aircraft as it did when the blades were out of balance. A lack of such tuning would cause the machine to jump from one foot to the other on its spider-like undercarriage.

"Pogo" had a short but not unproductive life with Atlas. It made some money towing advertising banners for the Central Canada Exhibition, Freiman's Department Store, and Jack Snow Jewelers. It also provided a good observation post for some search and rescue work. It was later sold, but destroyed in Toronto when a pilot failed to heed the unusual take-off procedure.

~~~~~~~

## Down the river...

Not long after Weldy was licensed for commercial operations he was sent by Atlas to the Lower St. Lawrence River area to fly in support of the forest industries of the area. That included hopping employees of two paper companies to various sites, or supporting forest surveys and fire patrols. Operations were flown out of Atlas bases on the north side of the river at Papinachois and Petit Bras near La Malbaie (known in those days as Murray Bay).

His days in the Lower St. Lawrence were far from boring. Being the manager of two satellite bases of an Ontario company in the province of Quebec posed challenges equal to the most exciting flying job. Across the 75 km of the St. Lawrence River, where a Quebec aviation company operated, there was resentment of the intrusion of an Ontario operator in Quebec. This boiled to the surface on occasion and threatened not only the safety of the operation but its very existence.

Weldy had loaded his Anson V with cargo and was rolling down the runway when smoke filled the cabin. Quickly he aborted the take-off and jumped out of the aircraft. From a cargo hatch he could see more smoke. He grabbed a fire extinguisher and put out the flames. It was quite evident that when the last item was placed in the hold, burning material was included with it.

Fire was the method of choice on another occasion. A local workman was hired to keep the hangar at Papinachois in good order. He was a man easily tempted by a free drink. And the drinks flowed freely one evening when nobody else from Atlas was around. The donors of the booze suggested that a nice big bonfire right beside the wooden hangar would be a good idea. And so the hangarman set a blaze that had nowhere to go but up the side of the hangar. But just as the flames got into the intended position, Weldy and Jimmy Lago drove onto the property. They quickly stopped the flames before the hangar caught.

Another incident did slow the operation somewhat. When Weldy was loading a Seabee on the south side of the river it was "accidentally" rammed by a truck. The tail was so badly damaged that it had to be replaced. The aircraft was stranded until Angus Morrison could drive from Ottawa with replacement parts.

In addition to the complaint that only Quebec companies should be allowed to operate in the area, there was a semi-legitimate claim that Atlas was carrying out scheduled services without a proper licence. It was true that scheduled flights moved back and forth across the river. The contracting paper companies flew lumberjacks out of the woods at regular intervals for off-time and back again to resume work. In response to complaints from competitors, Atlas pointed out to the Air Transport Board that it was the companies' schedules, not their own.

Such company flights were normally made in Anson V aircraft. The twin-engine Ansons were war surplus but in new condition. Creature comforts were not luxurious but the flights were short and the accommodation proved adequate. Canvas-seat benches ran down both sides of the cabin providing space for eight passengers. A ninth could ride up front with the pilot. Such non-sked flights did not require a co-pilot, but on one occasion an extra hand would have come in handy. A passenger who had too much time to kill waiting for the flight managed to get roaring drunk. Just as the aircraft was about to touch down, the passenger put a headlock on Weldy from behind and insisted that he have a swig from the bottle of rye he clutched in the other hand. Somehow the Anson found its way safely onto the runway as Weldy fought off the generous passenger.

As manager of the area operations for Atlas, Weldy did a first-class job against nearby competitors. But his skills did not go unrecognized by some of that competition. After a couple of years of resenting his presence in the area, Rimouski Airlines hired him as a pilot.

Airline flying was a dress-up job – uniform, shave, haircut and all that. His sister Gerry recalled how smart he looked in his uniform, an anomaly for Weldy who had turned the casual look into a fashion statement. The new job, however, must have appealed to him in the beginning as a job with greater prestige than hauling drunken lumberjacks out of the bush and returning them hung over a few days later. But the hike in social standing afforded by airline flying proved less than he expected, and he only stayed for one year.

Although he was reluctant to tell stories on himself, he enjoyed recounting tales of others. One of his favourite memories was waiting at Rimouski airport to take over the incoming Anson for the next leg of its route. As the aircraft taxied slowly up to the terminal, a small knot of passengers assembled to board the flight. Captain Phipps, turned out in his smart uniform, stood with them in the fresh air.

The Anson came to a stop in front of them, its two engines just ticking over. Then it performed a gentle kneeling manoeuvre, the wheels retracting slowly, the aircraft descending until its props hit the tarmac, coming to rest quietly on its belly in front of the astounded passengers.

The door flew open and a young pilot with three stripes on his sleeve burst out onto the waiting area. He was hotly pursued by an older pilot with four stripes on his cuff. With passengers at ringside the captain proceeded to beat the hell out of the co-pilot.

# Chapter 3
## Northern fever

After the Second World War Canadian political interests turned inward. Increased attention was directed to the under-explored northern regions. That interest was both political and economic – defence against the perceived threat from the Soviet Union, and the quest for mineral resources. The North American Air Defence (NORAD) agreement led to the establishment of the Distant Early Warning (DEW) Line in the far north, and its southern cousin the Mid-Canada Line. The intent was to detect hostile aircraft in the High Arctic before they could reach populated areas. The Pine Tree Line was established in more populated areas to direct interceptors against intruders.

In the face of increased American interest in the Arctic, Canada recognized a need to certify its sovereignty in the barren territory. There was a multi-faceted need to know how this unexplored frontier could be investigated, surveyed and mapped as well as defended. Canada's expertise, particularly in the area of mineral exploration, put a scientific label on the sovereignty initiative.

High-altitude photography to provide mapping at four miles to the inch had been undertaken by the Royal Canadian Air Force. Wartime bombers such as the Lancaster were deployed at a growing number of northern airports that were being established or upgraded in the postwar era. There was an increase in flights of transport aircraft to supply base camps during the summer months. Helicopters, which had to be staged north by way of a series of fuel caches, proved very costly to operate.

As the massive operation expanded, civilian companies were formed to survey and map the vacant landmass. Experienced pilots, navigators and mechanics were in good supply following the war.

Eager to make their mark in the unfamiliar peacetime economy, they put their expertise to work in the mapping business. One of the companies formed in the post-war aerial survey expansion was Spartan Air Services Limited based in Ottawa. Many veterans with similar military experience found their place in government regulatory positions. The new bureaucrats' familiarity with the technical discipline of air force operations tended to use military standards on the civilian operators. That made life difficult for those trying to fly, modify and maintain aircraft under frontier conditions.

Weldy Phipps joined Spartan in 1949 with duties as pilot and maintenance engineer. Within three years he became chief pilot and maintenance superintendent. He was made operations manager and head of the research and engineering division in 1954 and three years later, at age 35, he was appointed assistant manager.

By this time Spartan had 35 aircraft, including the Anson, P-38, Lancaster, Ventura, Canso, DC-3, Dragonfly Rapide, Super Cub, Husky, Mosquito and York. The latter two were acquired as a result of a shopping trip made to England by Weldy and Bill Law. The original objective had been to buy Mosquitos from Royal Air Force surplus, but during their absence the company's prospects for DEW Line work suddenly increased and so the shopping list was expanded to include several more mapping and transport aircraft.

"We were near Carlisle on the Scottish border to pick up the Mossies," Bill Law said. "We saw some other larger aircraft and asked what they were. They were Yorks. Since we had been ordered to pick up more aircraft, we bought 15 Mossies and 12 Yorks."

The Mosquitos had excellent qualifications for high-altitude and swift mapping operations. Keeping them fit for flight proved to be a different proposition for Spartan. Spare parts were not as easily acquired as they were for the surplus P-38s from the U.S. Repairs to the all-wood structure required careful measurement of the moisture content when using glue. Bill Law had just read about a new wonder glue called epoxy, but the aircraft were being phased out before much testing of the product, not yet approved by the DOT, could be carried out. Cost eventually phased the Mosquitos out of service. P-38 spares on the other hand were in abundant supply. Inhibited engines could be found in various locations. There were so many P-38s in Alaska that they had to be stored upright on their noses to fit them into available field storage space.

~~~~~~~

## The thinker thinks...

Weldy often lapsed into sessions of deep thought. His eyes, pouched and hooded from years of searching bright skies and snowscapes, concealed the mental machinations that pondered the next challenge, the next solution. It was a constant but obscure meditation on the decision that would advance his pursuit of excellence. Nobody really knew what was on his mind when these meditative states overtook him. Communication was suspended.

On one occasion, his three-year-old daughter Janet penetrated the stupor. Weldy had drifted off into one of his deeply pensive trances. Others in the family took little notice, but Janet, looking at her mysteriously spellbound father, said with some impatience: "Land, Daddy, land." He landed with a smile, but such flights recurred frequently and formed the genesis of great ideas and an instinctive reaction to the constant

The sleek beauty of the P-38 was blunted somewhat by Weldy's bulbous redesign of the nose but flight performance was actually enhanced.

Bill Law Photo

admonishment: "You can't do that, Weldy!" It was a reaction that got its noblest exposure with Spartan Air Services.

Any discussion of Weldy Phipps and Spartan Air Services usually brings up the war surplus Lockheed P-38 Lightnings used by the company for high-altitude mapping. The P-38s were actually the F-5 photo reconnaissance version of the American fighter. The fighter version had four .50 caliber machine guns and one 20mm cannon, with 150 cannon shells and 1600 machine gun rounds. It had a payload of two 550-pound bombs. Such a lethal load earned it the name the "fork-tailed devil" by the Luftwaffe.

Stripped of its armaments, the F-5 lost some of the sleek beauty of the P-38. Its unmodified nose, which contained three cameras, was blunt by comparison. The F-5 had been considered by Spartan to be the answer to high-altitude photography but proved less so in trials: there was no room for a navigator/camera operator. The task of solving this problem fell upon Spartan's maintenance superintendent and chief pilot, Weldy Phipps.

As was normal when such challenges arose, Weldy went into one of his deep meditations, the cogs of his brain meshing silently. And then with the quick command of a football coach ordering a change of play he said: "Cut 'em off right about here," indicating a position about two feet from the tip of the nose. His technicians gasped but began to cut the nose off four of the P-38s. John Roberts, Spartan president, entering the hangar the next morning, viewed the four aircraft with the nose surgery and uttered only one word: "Kee-rist!" There stood four of his operational aircraft brutalized well beyond serviceability. His worries, however, were short-lived. Weldy's natural ability for jury rigging and modifying aircraft was not yet understood, but confidence in his skills was enough to buffer the shock.

With the nose amputation completed he proceeded to splay the resulting cavity open a bit to make more room for the second crewman. An escape hatch was added, along with a bubble into which the navigator could insert his head to check for drift or make other necessary earthward observations.

Next came modification to the supercharging system for the engines. Weldy was not a lone wolf when it came to technical solutions. In the case of the turbo-supercharger, he consulted with the Lockheed company, explaining the turbulent air flow he had discovered. Lockheed pointed out that the P-38 was a combat aircraft and if flown under less stressful conditions would not require the vanes that he found in the duct work. Weldy's curiosity got the best of him. He removed one wing from a P-38 and examined its internal structure carefully, marvelling at the welding of the internal vanes from the outside of the duct. Then he removed the vanes and rerouted the duct work that ran through the wing roots at the fuselage. He installed a vane to direct a constant flow from the turbo in each boom.

Bill Law, by this time a graduate mechanical engineer, continued to be a party to many of Weldy's technological assaults on aircraft. For seven years he did all the stress analyses and drawings for the improvements. He recalls the informality of the planning phases of a new modification. "Many were the nights after working until late that Weldy, his brother Gordie and I would stop at the 'Last Chance' (a.k.a. Prescott Hotel) on Preston Street for a beer. It's amazing how many engineering details for the P-38s were worked out on the backs of cigarette packages or paper napkins. Certainly the nose escape hatch was."

The Phipps-Law team established a quasi-legal procedure for aircraft modification that can be credited for the company's success in mapping much of the Canadian hinterland. In the urgency to get the job done, the many government regulations in place to ensure aviation safety were routinely circumvented. "Oh Lord!" said maintenance engineer Bill Doherty, "If you asked for approval of this stuff before you did it you'd have never got it. There was no way you could wait to get approval and make a living. You had to go ahead and do it. We would never have got off the ground if every time you did a camera hatch or anything you asked for approval first. Bill Law drew up the plans and stress analysis – he was an engineer and he knew his business.

"We were away ahead of the guys in DOT but there was one thing you had to do – weight and balance. We used to put in some kind of mistake in our submission just to give the DOT something to correct. That made them feel useful. All the modifications were done and flown – often for the whole season – before they were ever submitted for approval.

"The modifications were all Weldy's ideas. Each year he'd come up with new ones after a season's operations. Could we change this or that? Weldy would just say 'Give me a couple of men.' We worked together all the time but he did all the mods."

When a change was physically in place, Bill Law would complete the drawings based on what Weldy had done. When the innovation was tested, usually in the field, the paper work would be taken to the Transport Department regional offices in Toronto and presented for approval. The normal reaction was: "Hmmm... let us know when you are ready to fly. We'll have to inspect the aircraft. Where is it located?" At this point Bill Law would clear his throat and, in a roundabout way, admit that the modified aircraft was out of town right now – for a couple more months. While this rankled the regulation-bound bureaucrats, they gradually learned that if Weldy Phipps had a hand in the work they might safely sign it off.

One memorable test flight sticks in Bill Law's mind. The big plastic bubble Weldy had installed on the nose of the P-38 required a stress analysis. While such a process would normally begin with mathematics and physics, Weldy decided there was a quicker way. And so Bill and Weldy took off in the modified aircraft and flew to 30,000 feet. Weldy then put the aircraft into a high-speed dive while Bill sat in the navigator's new cabin and held his hand against the perspex. He was to let Weldy know if there was any vibration. There was none and so the procedure moved on to drafting and eventual submission to authorities. The broader beam of the forward section and the relatively blunt plastic nose did not impair the P-38's performance. It actually improved the speed slightly.

Jean Greenfield, who began her eight-year stretch at Spartan in 1952, refers to the company as a family. "We worked long hours," she said, "but there were few complaints because we felt like partners. Often I would start at 9 a.m. and get out of the place at midnight. We had a lot of fun." Miss Greenfield started as secretary to president Johnnie Roberts but moved to the airport as head of the Operations Room.

Weldy's working habits, which knew no particular time restraints, evoked a response from some of the mechanics. Not particularly fond of the long hours, one recent arrival from England undertook to form a union. The idea failed to go anywhere, the family compact appearing to have squelched the plan. Compensation for the long and many extra hours came in the form of profit-sharing each year. However, that incentive was ended when the company went public.

There was an informality in much of the work around the hangar. Material was in short supply. "I remember Weldy and I scrounging around the hangar in nooks and crannies for nuts and bolts to fix aircraft," Bill Law recalls. And aircraft were not the only vehicles repaired in the hangar. In 1950 Weldy acquired his father's 1930 Dodge. Right away he overhauled it in the hangar, replacing all the main bearings, tightening them as hard as he could. When it came time to start the engine the starter would not turn it over. He got a tractor and pulled the car around the hangar, its rear wheels refusing to crank the engine at all. Finally the wheels turned and the engine came to life. Weldy let it run for an hour and it worked like a charm after that. He sold the car to Bill Law who ran it for some time before selling it to Jim Murray of the radio shop for $100.

~~~~~~

### Thin air for a moonlighter...

Back in the early fifties, Doug MacLeod toiled as an aircraft technician at RCAF Station Uplands. When his day's work was done he jumped the fence onto the civil side of the property and went to work for for Weldy Phipps at Spartan Air Services. "I'd work there until about midnight," he said., and Weldy would always be there. Weldy, in turn, often jumped the fence in the other direction for a beer in the airmen's canteen.

"One Sunday morning in 1951 I met Weldy after church at the Uplands hangar. We had worked late the night before upgrading a P38 Lightning that he had modified to carry a camera operator/navigator. in the nose section where the machine guns were in wartime.

"We climbed aboard and were in the air in a few minutes. I remember at 29,000 feet happily singing away. Weldy went on up to 35,000, ran off the engine and control tests and called me on the intercom. Of course there was no answer. He kept calling and repeated, 'Turn on your oxygen 100 per cent!'

"Weldy said later that . I had become unconscious from lack of oxygen, He told me that he had pointed the nose down, flaps down and speed brakes out and kept calling, 'Oxygen to 100 per cent!'

"He pulled out near 10,000 feet and I was able to answer him at about that altitude. My vision narrowed but I remember the flash of white concrete as we came in to land. I vomitted for two days. Still in the RCAF, I would not report my problem to anyone. I kept working at Spartan part time.

"Weldy saved my life by reacting quickly. I will always be grateful."

Doug Macleod went on to a full time job with Spartan in South America and later, as an employee of TransAir based in Resolute Bay, he was in contact with Weldy although Atlas and TransAir were competing for some of the work. He now lives in Red Deer, Alberta.

~~~~~

## A spiritual experience...

Buying surplus aircraft in the United States was not a very formal event. Weldy and engineers would visit the vendor, kick the tires and turn the engines over. A couple of revs of the engine and all the rust would be worn off the cylinder barrels, so why worry?

On one occasion in 1952 Weldy was ferrying a P-38 from Cedar Rapids, Iowa, to Ottawa when a wing tank booster pump failed. After landing for repairs, he removed an inspection panel. Instead of the usual array of wires and tubing he saw a lot of packing material inside the fuselage. He pulled the material out bit by bit and discovered eight bottles of liquor. Customs inspections were not a problem on such flights and so Len McHale and Bill Law felt it was safe to make some economical purchases in the States.

~~~~~~

## Some assembly required...

Spartan bought a P-38 from Kenting Aviation. Weldy, Bill Law and Maurice Giroux went to Oshawa to bring it back to Ottawa. The aircraft was in pieces and the crew had to put it together in order for Weldy to fly it. One carburetor was in a box, completely dismantled. With no manual to guide them they had to guess where all the pieces went. When the job was done the carburetor was installed and, to everybody's amazement, the engine started.

Weldy took off and headed for Ottawa. Upon approach, he performed a low slow roll over the Carp airport. Safely on the ground at Ottawa the rag tag aircraft was given a thorough inspection. The loose wheel fell right off the control column onto the cockpit floor.

~~~~~~

## A matter of survival...

Flying extensively over the uninhabited expanses of the sub-Arctic posed concerns should a crew be forced to bail out or crash land. To offer some protection Spartan fitted seat packs in each P-38. They contained a vacuum-packed sleeping bag, a .38 revolver and a supply of sealed air force rations.

While the survival pack offered psychological comfort, Weldy decided that it should be tested under winter conditions. He organized a crew consisting of himself, his brother Gordie and Bill Law. A Beaver aircraft dropped them onto a frozen lake north of Ottawa to see how difficult it might be to survive for 24 hours. The plane was to come back the next day and pick them up. The trio set up a tent, complete with sleeping bags in the middle of the lake. They had taken this gear along rather than ruin the sealed seat pack. The plan had been to live under a parachute canopy but it was only used as a partial windbreak.

When meal time came around the trio tried to open the rations. But they had been enclosed in heavy fabric and sealed with aircraft dope. No matter how they tried, they could not open the pack. This offered an opportunity to experience hunger, a condition Weldy knew well from his days as a POW. Anyway, they reasoned, they would be picked up the following day. When daylight arrived, however, they were greeted by a howling blizzard. With nothing to eat they got a fire going and just to practise cooking food, boiled some water for tea. On the third day they decided to walk out of their "crash site" towards some cottages on a distant shore. The parachute was abandoned as too difficult to carry in the wind. (When it was discovered in the bush years later it raised fears that some flyer was missing.)

As the trio struggled through thigh-deep snow, Weldy and Gordie took turns breaking trail. But the road was at least a half mile away. One of them discovered a package of air force high-energy "emergency jellies" in a pocket. Each survivor ate one and found new energy. "We just romped to the road," Bill Law recalls. They flagged down a ride into Low, Quebec and took a bus to Ottawa. Not much came out of the test except the addition of a parka·hood on the flying suit to keep the snow from getting inside the collar. A sharp device was added for opening the rations.

~~~~~~

Moulding Weldy's body for a pressure suit became a long ordeal. Chilled to the bone, he finally had to be rescued.

Bill Law photos

## Applying a bit of pressure

When Spartan started to fly at high altitudes, Weldy decided that a pressure suit would make life a lot more comfortable. Such a suit would be a skin-tight fit with chambers to inflate and squeeze blood back up to the brain. An existing suit was not snug enough for his build. As usual, Weldy felt competent to design and construct his own suit. The unusual tailoring job was undertaken by Bill Law and Maurice Giroux. The first step was to take a mold of Weldy's body by covering him from the neck down in plaster of Paris. Work began just after noon. He was wrapped like a mummy and painstakingly trowelled into the full-body cast, an exercise that took a quantity of beer in addition to plaster.

Work went on well into the night. But before the construction was completed, Weldy began to complain of cold. The curing wet plaster had begun to refrigerate his body. Finally he had to cancel the experiment, claiming to be nearly frozen. The suit builders, now joined by Bill and Mary Doherty had to extricate Weldy from the hard cast using heavy-duty tin snips, knives and any other sharp devices handy. Following that experience Weldy shopped at the RCAF's Institute of Aviation Medicine in Toronto, where the original pressure suit had been invented.

~~~~~~

### Venture in a Ventura...

High-altitude mapping required well-calibrated cameras, and calibrating them required some very precise navigation. That posed a temporary problem one day when no navigator was available. That called for one of Weldy's slap-dash solutions, which he was not long finding: he got Bill Law to go along with him in a Ventura bomber that he had converted for mapping. It was another first for Bill.

"When Weldy had the aircraft roughly set up on the camera run," Bill said, "he clambered into the nose to navigate and told me to get into the pilot's seat and follow his instructions. Up to then I had only flown Piper Cubs. This was like driving a truck without power steering. Anyway, the run was accomplished and Weldy came back to take over. Neither of us thought that anything could go wrong. All Weldy was concerned about was that my licence was valid."

~~~~~~

## A digit for delinquency...

In spite of his many "experiments" and misadventures, there is no record of Weldy Phipps ever receiving Spartan's Order of the Golden Finger. That was a light-hearted award conferred upon Spartanites who had done something really dumb.

~~~~~~~

## All this and fish, too...

In 1954 Weldy flew Russ Hall, senior partner of Spartan, and engineer John Deacon far into the Coppermine area of the Northwest Territories. Their mission was to find suitable terrain upon which to build an airstrip to serve Spartan's high-altitude mapping and installation of short-range navigation (Shoran) units.

They flew from Norman Wells north to the coast along the Coppermine River in a Fairchild Husky (CF-BQC). Several beach sites were looked over but Weldy rejected most because he felt the local weather would be too restrictive. Their attention was directed farther inland again where fog would be less of a hazard. At Pelly Lake, 1000 km northwest of Churchill, Manitoba, they discovered a long gravel esker, a geological feature deposited by receding glaciation during the last ice age. Reaching out along the shore, it was positioned ideally to the prevailing wind. It afforded tight but adequate length for the mapping aircraft and a ramp profile for beaching of the Canso aircraft.

In addition to the site's aeronautical suitability, the adjacent waters were teeming with fish, a major attraction as far as Russ Hall was concerned. He would haul in the Arctic char and Weldy would fry them on a small camp stove in the cabin of the Husky.

With all obvious features accepted, the trio set about recording some topographical data. John Deacon had a transit and took some measurements. Then they undertook to locate the area on a map. But the only map available at the time was on a 1:250,000 scale, or three miles to the inch, which made precise location a problem. But Weldy had a sextant and he taught the others how to take sun shots and resolve them to their exact position. "We all took our shots," John explained, "but we each came out at least three miles from each other. Weldy's measurements proved the only accurate ones."

With the Pelly Lake site chosen, Weldy had to fly 650 miles to Norman Wells for more fuel. He had to cache barrels of avgas along the route back to Pelly Lake in order to supply the site. As he departed, Russ Hall remembered never having felt so lonely. Locating such a tiny lake again among the hundreds that dotted the small-scale map would require expert navigation and perhaps a lot of luck. Weldy might run into disaster and leave them there forever. But he returned a few days later to pick up his passengers and begin preparations for the establishment of the airstrip.

The next step was to dispatch the Husky from Norman Wells to Pelly Lake with a heavy load of equipment. The generators, tents, batteries, and gasoline proved too heavy for the aircraft and had to be off-loaded progressively until it could get airborne. At Pelly Lake, unloading became an unwieldy task because docking facilities had not yet been built. The generators weighed about 200 pounds, and drums of gas to power them were not light. John Deacon, civil engineer on the project, recalls: "In order to get the cargo ashore we had to stand waist deep in ice cold water and pass the goods along."

When they had unloaded the Husky and prepared to set up the camp they looked up to see the aircraft drifting away from shore. Al Macnutt, the pilot, could not swim and so John jumped into the frigid waters and swam to the departing Husky. He clambered up onto the float and got into the cockpit. He was not a pilot but he had seen Al start the engine often enough. He engaged the energizer and turned the engine over and over. No start. He tried again and still the engine failed to start. The aircraft moved farther and farther from shore. On the third try, with an overdose of gas in the cylinders, the engine burst into life sending out a huge belch of smoke.

John felt a bit more confident now and began to taxi the aircraft up and down the shore. He could see Al running along with him, obviously highly excited. Eventually John moved the aircraft back to shore where it could be secured. Knowing the aircraft's habits so well, Al had been convinced that the aircraft would catch fire in the hands of its accidental pilot.

~~~~~~

### Hacksaw to the Rescue...

When the new airstrip was finally in place at Pelly Lake a Spartan DC-3 landed on the runway. All was well until it taxied onto a patch of soft sand near the camp. There it sank nose down, its props hitting the ground and bending the tips.

The damage was reported to Weldy who flew to Pelly to take charge of the situation. When the aircraft was jacked out of the sand he set about trimming the ends off the props. Under normal circumstances the balancing of props is a highly precise machine- shop proposition, but Weldy was already well known for his makeshift technical skills. He cut and filed the tips until he judged their vibrations to be safe enough for flight. And then he flew the aircraft the 1000 km to Churchill to wait for new props.

Some years after the establishment of the Pelly Lake airstrip, Weldy got a hankering to return in winter. With John Deacon and Vic Koby, then editor of Canadian Aviation magazine, he flew an Anson to the strip. He assumed correctly that the snow would be firm enough to support the Anson. The trio stayed overnight in a camp building and departed the next day after heating the engines with blowpots. Failure to get the engines started would have meant disaster.

Front-line airstrips were an essential part of the Spartan operation. In addition to the strip at Pelly Lake, 320 km. northwest of Baker Lake, N.W.T., Weldy supervised the construction of airports another one at Esker lake in the Ungava District. Borrowing some earth moving equipment at Dawson City, he made great improvements on a gravel strip there.

~~~~~~

## One wing wonder...

As chief pilot, Weldy enjoyed the private use of one P-38 (CF-HDI). It had not been brutally modified for aerial survey work; its only civil conversion was an uncomfortable observer's position up behind the pilot. This seat allowed a converting pilot to see how the aircraft flew, thus preparing him for his own turn at the controls. There was no hands-on equipment for the passenger.

The year was 1953. Another Spartan P-38 approached Dawson City, Yukon, on final for a landing at the Bonanza Creek strip. An engineer on the ground waved it off because one of its wheels was not down. Unable to get the gear sorted out, the pilot broke off and flew to Whitehorse, where facilities were more accommodating for an emergency landing. Along the way the drop tanks were jettisoned into Lake Labarge (where strange sights had already been seen, thanks to poet Robert Service).

The landing was noisy and exciting for the navigator, Bob Bolivar, seated in the nose of the aircraft. But there was no fire, and except for a severely scraped belly the aircraft suffered only repairable damage. An air force crane was engaged to lift the aircraft onto a flatbed truck. It was raised carefully above the truck. Then for reasons unknown, it suddenly dropped. The starboard wing broke off on impact.

Weldy was called to the rescue. He flew from Ottawa in HDI to assess the situation. Unwittingly, he had brought the major component required for the repairs with him: the right wing from his personal P-38. He had it removed from the functioning aircraft and transplanted onto the damaged one. All local supplies of fibreglass were bought up in order to reattach the turbocharger ducts. In short order the operation was back on line. Another wing was ordered for HDI.

~~~~~~

CF-WWP ended up in the pond at Pond Inlet when a tire blew on landing. Weldy enjoys a refreshing beer while waiting for a replacement to arrive from Resolute.

A more messy situation befell the Twin Otter when the landing area looked safe but turned out to be deep gumbo.

After the rigors of the Arctic, Weldy and Fran relax in the sun at Lacaya in the Bahamas.

Fran describes this 1970 photo of her husband as 'Weldy in his element'. He made up for years of harsh Arctic temperatures by retiring to tropical seas on his 48-foot yawl Whiskey Papa.

Phipps collection

A Carribean Christmas aboard Whiskey Papa was snug and snowless. Fran, Terry's husband Brian Bean, Jim and Weldy gather around a pile of presents. (Note Weldy's partially-built Super Cub model overhead.)

Jim and Bob's base tents on Axel Heiberg Island had to be held down by fuel drums on each corner. They were stranded there alone during a storm.

Weldy and son Jim pay their respects at the graves of Franklin expedition members on Beechey Island, N.W.T. On other visits artifacts and messages were found. A cache of tinned beef tasted good after 100 years buried in the permafrost. Pilots ate 10 pounds of it.

Phipps collection

A man for all reasons, Weldy flew a variety of aircraft in his career. His favourites were deHavilland types but his early days included the Stearman and the Piper Super Cub. Note the overload of gas cans bound for a rescue of a search and rescue Canso north of Alert.

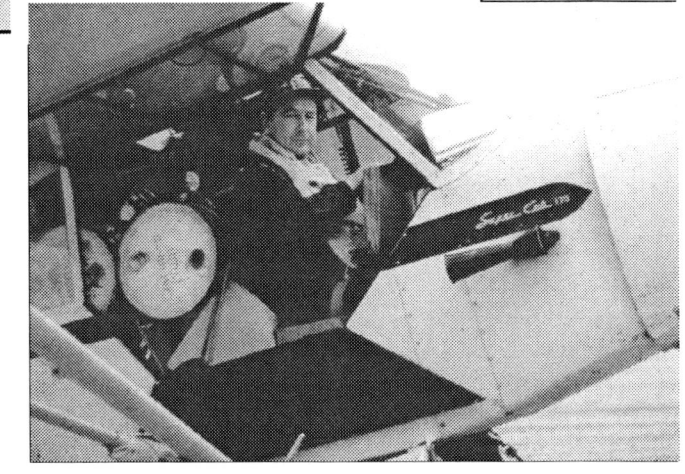

The dressed-down look was more to Weldy's liking than more formal garb. But he was smarly turned out to receive his Order of Canada from Governor General Jules Leger. Aboard Whiskey Papa, he was more at home in his heavy weather gear.

## Water, water everywhere...

All the airmanship in the business is nothing without a heavy dose of good luck. Such was the case on one flight from Pelly Lake to Churchill in a Spartan Mosquito. As Weldy proceeded along the 1000 km route he received a report that the ceiling had dropped to zero, eliminating any chance of a safe landing at Churchill. He decided to try and land at the small airstrip near the town of Gillam, Manitoba, 150 km from his destination. The weather was bad there, too, but he calculated that he had fuel enough to stay in the area to await any break in the fog.

He radioed ahead to Churchill to see if somebody could light a flare path at Gillam. Perhaps he could distinguish the burning pots of oil through the haze and get in safely. His ground contact alerted a railway worker at Gillam who knew what a flare path was. Knowing that the fires were set, Weldy began to let the Mossie down through the low overcast. As he came close to the ground he caught sight of the flare path and prepared for the landing which he knew would be critical because of the short runway. Just as he was about to round out his approach for touchdown he got his first glimpse of the landing area – open water! The railwayman had thought that a float-equipped aircraft was coming in, and lined one shore with flare pots.

Weldy quickly slammed on power and climbed away from the water and circled the area. Then, committed to a landing at Gillam, he descended through the overcast again until he could see the semaphore signal lamps on the railway. He knew that the airstrip ran parallel to the Churchill rail line. A few feet above the ground he shifted the course of the aircraft over to where he estimated the strip should be. This time he flared out right over the button of the short runway and landed safely in near-zero visibility. Getting the high-performance Mosquito off the runway the next day was almost as daunting a task.

~~~~~~

## Where's the fire??...

Getting a Mosquito off Carp's short runway one day presented another sort of problem. As Weldy roared down the runway and lifted off, vibrations tripped the sensors that controlled the system designed to put out any fire the aircraft might experience in a crash. In this case, the devices blasted fire-suppressing gas into both engines at the same time, snuffing out all power. With a crash seemingly inevitable, the windmilling props refired both engines.

~~~~~~

## Mean and lean...

Weldy once made a trip to South America to check on Spartan's operation of Mosquito survey aircraft in Colombia. While there, he and pilot Al Macnutt decided to travel across the Andes to visit Rocky Laroche, another Spartan employee, at Villavacencio. For some reason they journeyed back across the mountains at night by taxi.

"It's a precarious trip over the hill even in daylight," Al recalled, "but reinforced with *Agua D'iente*, we were sedated somewhat. Making a sharp turn in the hills, the cab driver lost control. The car skidded to a stop, balancing on its belly with the front wheels in space over the edge of a cliff. The driver, thinking the car would plunge down the mountain, scrambled through the back seat and out the rear door to solid ground. He knew how far down was, but we didn't.

"Cars going over the mountain pass climb from approximately 8500 feet at Bogota to about 13,500 feet. To get performance out of the engine the drivers get out half way up the hill and adjust the needle valve on the carburetor to compensate for the thinner air and keep the mixture at a reasonably correct ratio.

"Weldy was very interested in this local custom and was trying to explain to the driver, who spoke '*no Ingles*', how to install an aircraft type of mixture control that he could adjust from the inside of the car. His mechanical mind never relaxed."

On another southern inspection jaunt, Weldy visited the crew in Venezuela, where Al Macnutt and Hal Mordy were working on a geographical survey. "Weldy had not spent time in South America and, unlike most people, didn't want to see the bright lights. He wanted to go prospecting." Al and Weldy rented a Helio-Courier from the local charter company and flew into the interior to a tiny strip on the Orinoco and hired a guide to take them looking for gold. They spent a week camping at night and prospecting by day. "He was a different man away from the pressure of flight operations." Al said.

~~~~~~

### The Arctic Queen...

In the first post-war decade the quest for resources and scientific data in the High Arctic drew the enthusiasm of a variety of researchers – wildlife biologists, botanists, archaeologists, and an engineering entrepreneur named Dr. George Jacobsen, president of the Tower Company, a resource and construction organization. With visions of construction opportunities he formed an expedition to the north in late July, 1953.

The group included Dr. Erling Porsild, Dominion Botanist from the Museum of Natural Sciences, veteran Arctic explorer Dr. Graham Rowley, and Dalton Muir, wildlife biologist and National Film Board science filmmaker. In August, the group boarded a Canso aircraft called The Arctic Queen, piloted by

Weldy in the cockpit of the Canso that explored the High Arctic and found a "warm" lake where others were frozen.

--Dalton Muir photo

Weldy Phipps. His co-pilot was Frank Pynn and his navigator was Jock Buchanan, a couple that had crewed on the same RAF bomber during the war. The maintenance engineer was Norm Hineson and Jim Murray was the radio operator. They droned northward, and finally landed and set up camp on Axel Heiberg Island on the shore of a lake that commanded much of the press attention on their return. The freshwater lake, nestled between two rows of mountains, was free of ice while water for 800 km to the south was still solid ice.

The "warm water" phenomenon was eventually explained by the venturi effect which accelerated the already high winds down the lake between mountain ranges. The strong winds scoured sand and dust from the nearby hills, darkening the ice surface. That attracted and trapped heat from the sun making the ice vulnerable to the force of the wind.

The expedition provided a learning experience for all hands, but it suffered from a subtle strain between the crew and an assertive George Jacobsen. As financier of the trip Dr. Jacobsen tended to intrude into operational and safety related decisions. He had even insisted that the Canso be named 'Arctic Queen', rather than its original 'Shrimp Boat'. Weldy was reluctant to become equally assertive in exercising his responsibility over flight decisions.

Frank Pynn and Jock Buchanan resented their perceived treatment as lackeys by Jacobsen, who apparently was unaware that he was interfering with some of their primary duties. However, Pynn had an enormous sense of humour, which tended to cool the situation. His favourite ploy was to pull Jacobsen's leg by telling exaggerated horror stories about crashes. "Canso aircraft burn very fast," he would tell him, "and a man has never been thrown clear of the wreckage." That tended to discourage the leader from urging unsafe air operations. Weldy let such tactics take care of matters and joined the fun by wearing a sign Pynn pasted on his jacket saying: I LOVE GEORGE.

As the scientists went about their particular studies they became fascinated by a nearby ice cap and decided to hike to it. The topography of the High Arctic, however, can be deceiving. The group reached the ice cap after an unexpected 15-mile trek and still had to walk back. Provision not having been made for such a long period, the return trip grew more burdensome with each mile: they were hungry and exhausted. Back at the camp Weldy realized what had happened and met the group part way back. Dalton Muir recalled how welcome he was, bearing a hot thermos of coffee laced with rum. The trekkers had subsisted on one orange and some high-energy lemon glucose candies.

One situation not well understood by the scientists while they were deeply into their investigations was the rapid approach of winter. Jim Murray, already concerned about sporadic sun spot interference with radio signals, recalls waking one day and recording a temperature well below freezing. Contact with Resolute confirmed the likelihood that the cold weather had arrived for the winter.

For Weldy, the decision became critical. Weather authorities recommended a wait for the lake to freeze enough to allow a wheel take-off from the surface. Such a decision would prolong the stay at the lake until mid-November, an unattractive prospect for the party. A flying boat take-off could ice the props and wings of the Canso, making it impossible to hurdle the 3000-foot mountains. Weldy did not want to be marooned with his aircraft frozen into the lake and made quick preparations to get it out of the water.

"Weldy did a masterful job," said Jim Murray. "The operation had to be carried out with the engines since the beach sand was too soft to permit manhandling. It had to be right on the first try since our Canso did not have reversing propellers. There was no backing up and the cliff was very close to the beach. However, Weldy accomplished the maneuver with several feet to spare between the port wing and the cliff. We all breathed a great sigh of relief."

Fortunately, the weather warmed up. Weldy then announced that the Canso was leaving. The scientists, reluctant to abandon their studies, boarded the plane for the trip back south. Somewhere between Resolute Bay and Churchill Jim Murray received a radio message that he would not reveal to Weldy while they were airborne. It reported the death in a P-38 crash at Dawson City of Jimmy Lago, Weldy's wartime bomber pilot.

"Weldy never talked much about his wartime years," Jim said, "but Jim Lago used to fill us in on the details of their capture and the subsequent POW activities. News of his death was a terrible experience for Weldy. It was the first time I ever saw him show emotion. He and Jim were very close."

The Arctic Queen expedition helped to ignite the subsequent resource play in the Arctic. While George Jacobsen's primary interest was in Arctic construction, it was the discovery of two massive gypsum domes (uneconomical for recovery at the time) that demonstrated the potential for petroleum deposits in the region. Oil was found much later only 80 km away. Oil exploration enjoyed a brief period of activity which was overtaken by discoveries in more accessible areas. Weldy acquired resource exploration licences in 1959 for 67,000 acres on South Melville Island, but none paid off.

Dr. Erling Porsild's botanical investigations ultimately established the classification and origin of the Arctic flora and confirmed the circumpolar distribution of many of its species. Dalton Muir's evaluation of the potential environmental impact of resource development was recorded in reports and on film.

The Canso left its mark on Arctic aviation. The aircraft was later flown to Scotland where it became the victim of a vicious storm: it broke its moorings and was being carried out to sea. Its navigator, Jock Buchanan, jumped aboard the aircraft at the last moment but was unable to start the engines. The Arctic Queen was soon swamped and lost at sea. Jock Buchanan lost his life in the heroic effort to save his aircraft. The 'warm' lake on Axel Heiberg Island was later named Buchanan Lake in honour of the courageous navigator whose skill had located the remote lake in 1953.

~~~~~~

**Heads – you lose!**

Life in the hinterland included a lot of boredom, be it hours of flying precise mapping lines, magnetometer grids, or just sitting out weather or aircraft snags. During such a lull during the summer of 1951 some Spartan crews were grounded in Dawson City, Yukon. The time was being whiled away at the bar in the Pearl Harbour Hotel. Weldy was not quite in the same mood as the others as he sank into one of his mental ruminations. The solution, as they saw it, was to goad him into the spirit of the occasion.

It started with a bit of ribbing about the blonde Weldy had hidden in his room. Reaction was slow at the start, but continued accusations about the lady upstairs began to get under his skin. The more he denied, the more the boys bugged him. His temper rose, quite to the entertainment of his buddies.

The group, which included Bob Fowler, Len McHale and Bill Doherty, demanded to be shown the room. Weldy now was angry. He led the others up the stairs, and yanked open the door to his room. There, draped across his chair, lay silk stockings and ladies' underwear. "Aha! aha!" chorused the trio, sending Weldy into an uncharacteristic rage. In the bed lay a prone female body, face to the wall, her long blonde hair flowing over the blanket.

Weldy charged across the room and poked the form in the bed, ordering her "…to hell outa here!" When she did not react he gave her hair a yank, neatly decapitating the mannequin, stashed there by the hotel's owners, Rick and Kay Carswell. All survived the incident.

~~~~~~

**Cooling the Yukon...**

The Spartan crews worked long hours and were quite used to expending energy. When down time came along they tended to fidget a lot and were happier working at almost anything. Such was the case on one occasion when weather grounded the aircraft in Dawson City.

Bill Doherty recalls: "We decided to help John Donald who ran the Northern Commercial store. We were painting and so on and he asked if we could hook up his new freezers. We figured there'd be no problem and Weldy took on the job. The trouble was that the old freezers had to be taken out. Weldy decided that he'd go under the floor and just cut the pipes. He cut the pipes all right. He had his oxygen mask on but the ammonia came out and settled about an inch above the wooden floor. John had to close the store for two days. Finally we hosed it all down and got the freezers installed. Some of the jobs we took on were a lot bigger than we thought."

~~~~~~

## Rub- a-dub-dub...

The Spartan aircrews were not restricted to airborne transportation. There was the time (circa 1954) they were staying at the Ingraham Hotel in Yellowknife. Grubby from sweating around the aircraft, they looked forward to a refreshing bath. The bathtubs were free-standing metal vessels – no plumbing or feet attached. One could just sit and soak. Or – after rounds of overproof rum – get aboard one or more at the top of the stairs and, with a vigorous push, race down into the lobby like a luge rider.

The hotel management failed to see the entertainment value in this new sport and chucked the whole contingent out.

~~~~~~

## If you can't lick 'em...

It is an accepted fact that travel and ice cream will broaden one – albeit in different ways. But if you combine travel, ice cream and beer, it can be shown that there is also a way to be reduced. The travel was to Manitoba in 1955 during the aerial survey heyday of Spartan Air Services. Weldy was between flights on a hot summer's day in Churchill. To cool things off a bit he bought an ice cream cone. But just around the corner was a bar, and the thought of a tall, frosty beer seemed like an even cooler idea.

When he entered the bar he was informed crisply that "You can't come in here with that!" The waiter pointed at the ice cream cone.

"Why not?" Weldy asked.

"Food is not allowed where alcohol is served!"

Weldy licked the ice cream at an accelerated pace in order to conform with this unfamiliar rule. When the waiter became more forceful, Weldy began to laugh and ridicule the law. The waiter called the police who just as promptly arrested Weldy. He was fined $50. For years the receipt was an artifact in the Phipps home, giving Weldy his money's worth from the lingering souvenir of his crime.

~~~~~~~

## Chapter 4
## Bradley Air Services

By the end of the fifties, the heyday of the aerial survey business was in decline. Weldy Phipps, who had joined Spartan in 1949, left after 10 years to become a partner with Russ Bradley in Bradley Air Services of Carp, Ontario. He served as vice-president and chief of operations. This marked another significant milestone in his career as a visionary in the development of aircraft. During his tenure at Bradley he began to work on the use of light aircraft in the High Arctic – work that finally focused wider attention on his accumulated aviation achievements.

Having operated larger survey aircraft from crude airstrips in the sub-Arctic, he looked for ways to extend the use of aircraft that would not require massive support facilities in the far north. He first chose to equip a Piper Super Cub with a tandem Whitaker landing gear. That configuration had two in-line wheels on each side of the aircraft. He experimented with a caterpillar-type track over these wheels. The tractor-like configuration performed well. The Cub could climb snow banks and operate in fields of high grass. But the modification was not without serious problems: the track was prone to coming off in tight taxiing turns. A flat tire could spell disaster if the track flew into the prop, explained Bill Law, who nursed Weldy through such innovations from an engineering standpoint. The tandem configuration without the tread, however, was retained for one test in the High Arctic.

~~~~~~

### Rescuing Search and Rescue...

It was a sort of David and Goliath situation in the spring of 1956 when Weldy set out on the 3000 km flight from Ottawa to Resolute Bay in a Super Cub. It was not an undertaking that most pilots would choose without trepidation, but Weldy

OK, smartass, what was that crack about search and rescue?! After dire warnings that the air force would not search for him, Weldy had to rescue the search and rescue Canso north of Alert.

—Phipps collection

smelled a source of contracts for the light aircraft. That was enough to get him pointed towards the High Arctic in such an unlikely vehicle.

While he saw little to hinder his dream, the authorities got a case of the jitters. He was warned that he could not venture into such a vast territory with a light aircraft. "Why not?" The warning was tempered to "should not". The reason was that the air force would never find him once he got lost. And, of course, he would get lost. In fact, they added, they would not even search for anybody on such a foolhardy mission.

When he arrived in Resolute, he was again asked why he would venture so far north in such a small plane. And again the matter of search and rescue over such vast territory was cited. But as luck would have it, search and rescue was indeed an important element of the story. One of the first people he met at Resolute was a Royal Canadian Air Force officer whose duty it was to perform search-and-rescue operations. The Squadron Leader asked Weldy if he was available for charter. Why would the Air Force want my services? Weldy asked himself. The officer explained that he had ditched his search and rescue Canso 20 km north of Alert. He had run out of fuel and one wheel was damaged in the dead stick landing. "They needed my little Super Cub to rescue this search and rescue aircraft!"

The problem: how to carry the amount of fuel the Canso would need to get it back into the air? Weldy flew to Alert, a matter of 700 miles, to tackle the assignment. To make way for a

45-gallon drum of fuel, he removed the single passenger seat from the Super Cub. Then he set up a shuttle service flying several drums of avgas to the Canso.

* * * *

## The Fort Conger visit...

In August of the same year, Weldy ran into Dalton Muir, a wildlife biologist and filmmaker whom he knew from the Arctic Queen expedition. Dalton was on Ellesmere Island on a National Film Board photographic shoot. Weldy was quick to ask if Dalton could use an aircraft in his travels. Such a job would provide a good test for the Super Cub.

"My budget could never afford the cost," Dalton told him. Weldy offered a loss-leader bargain: "You pay for the gas and insurance and that will be all I want." The Cub set out across the top end of the island.

On approach to a river outwash, about 100 miles south of Ellesmere's northern tip, Weldy made a careful assessment of the surface. It looked smooth enough for an easy landing with the Super Cub, but it was a dull day and ground features were difficult to distinguish. Unseen from the air were a series of narrow washouts in the surface. On touchdown, the aircraft immediately dropped one set of the tandem wheels into one of the gaps and went into a groundloop. The Cub was seriously damaged: the A-frame of the landing gear was broken, a wing tip and the prop were clipped. There was damage to the rear of the fuselage. Also, the radio was out of commission. More embarrassed than otherwise concerned, Weldy did not want Dalton to photograph the damaged aircraft. He said that it was the first time he had damaged an aircraft in peacetime.

"We'll have to tell somebody where we are," Weldy said, recalling the dire warnings from the air force, "or they'll send out a search plane." (He didn't believe the no-search threat.) Weldy took some sun shots and established the position. The location of the accident was Fort Conger on Lady Franklin Bay. Before he could advise the base at Alert of their whereabouts, however, he needed the radio. Fortunately, he had a hand in the original design of the high-frequency radio, ensuring that its five crystal-controlled frequencies had good range. To repair the unit he removed all the vacuum tubes and replaced them with spares. But the HF radio still required a long antenna. Normally, the Cub flew a drogue antenna from the rear of the fuselage. To hold it in a horizontal position in flight a small plastic funnel was attached to the end. Before landing it could be reeled into the aircraft.

He had anticipated situations in which the 300-foot trailing antenna would be unavailable. As a precaution he carried a weather balloon which he filled from a small hydrogen generator. The balloon was then raised on the antenna wire about 50 feet above the aircraft and adjusted in length to match the proper frequency. This allowed contact with Alert. And in preparation for repairs, if and when they arrived at Alert, Weldy asked them to order some aircraft fabric and clear dope from the U.S. Air Force in Thule, Greenland. It was not reassuring to be asked: "What's clear dope?"

Haywire isn't found in most hangar tool cribs, but you could never bet on it when Weldy was on the job. His history of modifying both sophisticated and simple aircraft, and stories of makeshift en route repairs, testify to his philosophy that there's more than one way to skin a Cub.

In this case he fell back on his natural mechanical talent, training and experience. For reasons he probably hadn't yet established he had packed some flexible cable in his tool kit. This was used to lash the damaged landing gear back into place. It was reinforced with angle iron salvaged from a bed frame found amid the ruins of Greely's original scientific base, established in 1881.

Adequate repairs were made to other parts of the airframe. The tips of the prop were damaged but considered airworthy for the 150-km flight to Alert. A barrel hoop served as a runway grading device with which Dalton levelled the rough spots where the aircraft had come to grief. It only remained for the two men to grasp the wing tips and twist in opposite directions. This brought the wings into a position of reasonable horizontal balance. The plane now looked like it might fly once again.

When it came time to take off, Weldy could not distinguish the dangerous surface from terrain that would support the jury-rigged wheel. The fragile condition of the undercarriage could not tolerate even the slightest avoidable bump. Since Dalton had graded the primitive runway he knew where the small fissures were and where the ground was safe. He outlined a suitable strip by placing bleached muskox bones along the strip as markers. The aircraft wobbled into the air – like the Flight of the Phoenix – and set a course for Alert. On arrival, Weldy borrowed welding equipment to make proper repairs to the landing gear and patch the fabric with clear dope so unfamiliar to the military at Thule. More suitable materials were salvaged from two Lancasters that had crashed at Alert. Tubular steel was carted back to a shop and assembled to fit the calculations Weldy had made on the Super Cub. He had to work night and day and go through several cut-and-try tests before the aircraft was again fit for service.

That was not the first time Fort Conger had treated visitors with a lack of hospitality. The fort served as Greely's base in 1881 when 27 of his party were stationed there. But the ship carrying new supplies did not reach them as scheduled. In fact it was three years late; by the time it arrived, 20 of the men had died of starvation. One was shot for stealing from the larder, which had

Weldy's early work with Super Cubs in the north was during his partnership in Bradley Air Services. Note the "tarmac" of the High Arctic.

--Phipps collection

contained mainly dried lichens and seaweed. Seven were rescued from the ice. The body with the bullet hole was discovered years later. It bore evidence of cannibalism.

(Peary also used Fort Conger in 1899 after a trek of 400 km from his stricken ship.)

~~~~~~~

## Enter: Tundra Tires...

The Ellesmere Island trial showed Weldy the limits of the tandem gear on Arctic terrain. As well as the hazards of dropping the narrow tires in hidden cracks in the soil, the aircraft could not be turned sharply under its own power. A 180-degree turn on the ground required the aircraft to be picked up by the tail and swung around by hand. That proved difficult for one person because the four wheels would not act as a suitable pivot point.

Weldy began, in his customary fashion, to ponder the difficulties and possible solution to light aircraft handling on hostile terrain. The major problem was in the landing gear and so he concentrated on the tires. He first tried 25-inch tires on a Super Cub.. But that size was soon expanded to 35 inches (a size suitable for a DC-3) fitted on hubs from a Stinson SR5 Reliant. Various tires were tested but found to be too heavily plied and too heavy for the Super Cub. With engineering assistance from Bill Law the laminations were reduced to four. The weight now was acceptable. A stress analysis had to determine their ability to withstand punishment from sharp stones and heavy landings and still provide acceptable ground handling. The eventual tire was manufactured by Goodyear in Akron, Ohio. Minimum order: 30 tires. They didn't have fancy sidewalls but each was embossed with the title Phipps Special.

The Phipps Special tires were eventually extended to the deHavilland line of Beaver, Otter, Twin Otter, Caribou and Buffalo aircraft. The U.S. Army had plans to equip its Caribou aircraft in Viet Nam with the Phipps tires. For some reason, believed to be rivalry between the U.S. Air Force and Army, the tires went to Viet Nam and came back, never having been used.

To ensure product security Weldy had farmed out the manufacture of the military wheels to various producers. He had the molds for the wheels made in one place, cast them in another and finished them in a third. The molds were hidden, along with information on the wheels, at the Ottawa Feather and Mattress Company. He wanted to make sure that the stocks or the ability to reproduce them were not easily available to competitors without compensation. At one time the Phipps home in the west end of Ottawa had a supply of tires in the living room. There was no place else to store them.

~~~~~~

## Bermuda triangle off course...

Flushed with the success of his first flight in a Phipps Special Super Cub to the most northerly tip of North America, Weldy felt it was time to let Piper Aircraft in on the success of their plane. To impress Piper with the vastness of the north he borrowed a map of Canada and photographed a track from Pelly Lake, NWT, to Cape Columbia at the top of Ellesmere Island. Then Weldy and Bill Law set out in an Anson for Lock Haven, Pennsylvania, home of Piper Aircraft. On clearing U.S. Customs they were given less than encouraging advice that their route passed through the "graveyard of aircraft", an area called "Hell Stretch" by airmail pilots of the 1920s. On schedule, one engine died as soon as they entered the area. The fuel pump had failed. Bill spent the rest of the trip on the wobble pump supplying the engine with gas manually.

While the aircraft was laid up the two were offered a flight home by a local pilot. En route, radio contact with Canada confirmed no need to go through Customs and they proceeded directly to Ottawa. But it became evident to them that the hospitable pilot had a bad case of the DTs – so bad in fact that Weldy had to take control of the aircraft to get them landed. Weldy bought him a coffee at the terminal but he was shaking so badly that he couldn't drink a drop. Later they saw the aircraft taxi out for the return flight to the States via Hell Stretch.

~~~~~~

## The coming of age...

On November 5, 1960, the RCMP near Grise Fiord on Ellesmere Island flashed an urgent radio message to Ottawa. The community of 200 Inuit was in the grip of a serious whooping cough epidemic. A seven-month-old baby had already died, two more were in critical condition. The population was in panic. Efforts to get the United States Air Force and the Royal Canadian Air Force to deliver medical aid by any means had been unsuccessful. A conventional aircraft could not even be considered since there was no landing strip and the North was in its winter darkness.

The message set in motion a desperate search for a solution. Ottawa authorities didn't have far to go. Weldy Phipps got a 2 am phone call at his Ottawa home. The Health Department wanted to know if he could fly a doctor from Resolute Bay the 400 kilometres to Grise Fiord. At home in bed Weldy was more than 3000 kilometres from Resolute. But it was a rhetorical question: the RCMP, RCAF and other government authorities knew that Weldy Phipps was a man who could solve flying problems that nobody else could or would touch.

The only aircraft he could use was one of his Phipps-modified PA-18 Super Cubs. The lightweight balloon tires would permit a landing on the most hostile pack ice. But the Cub was stored for the winter at Resolute Bay. Weldy said he'd try. The RCAF was standing by to fly him to Resolute. He roused Erwin Keller, one of his pilots, and recruited him for technical support. Then the two boarded the RCAF Dakota for the long flight north.

Reaching Resolute was the easy part of the mission, despite a diversion for repairs to the Dak en route. The aircraft that was to make history had been stored with its wings and tail removed so that it would not occupy any more hangar space than absolutely necessary. Resolute hangar facilities were, to say the least, spartan.

But Weldy and Erwin went to work in the minus-30 temperature – Fahrenheit or Celsius it's roughly the same at that level.

Four hours later, they had the basic necessities of flight in place. The engine came to life. Weldy put Dr. A. H. Stevens in the back seat, waved goodbye to Erwin, and took off for Grise Fiord. The next challenge was finding the way. Navigation by any conventional means was out of the question. Visual navigation was impossible in the darkness. The magnetic compass was useless so close to the magnetic north pole. The bearings of the gyro compass and artificial horizon were frozen solidly in their summertime lubrication.

The answer came from above: the RCAF Dakota flew the route within visual range above the Cub, maintaining radio contact. Weldy's main concern was that fog might separate them, leaving him unable to find his way. The most likely prospect for the pilot and doctor in such a predicament would be disaster – freezing to death somewhere in the bleak Arctic. It was a prospect that was becoming too close for the Cub's occupants even without the loss of the Dak – the Cub's heater failed en route and the two men were already nearly frozen.

The RCMP at Grise had set up a flare path on the pack ice in the only area where the light plane could possibly land. And after more than three hours of flying Weldy spotted a fuzzy glow from the cans of burning gasoline. His view was obscured by a bank of ice crystal fog that had settled on the area. The windscreen had frosted up, permitting only a little peep hole. But he was committed to land on several counts. He had to refuel, he had to keep the doctor and himself from freezing to death, and, of course, the residents of Grise had to be attended to. And so, with characteristic courage and skill, Weldy let down through the fog and bounced to a landing on the rough pack ice, his innovative tires absorbing shocks that would demolish any more conventional aircraft. Dr. Stevens was extracted from the frozen aircraft and put behind a dog team for a six-mile dash to the stricken community.

Over the next three days the heroes of the mission had their work cut out for them. While the doctor performed his medical duties, Weldy had to keep the engine of the Cub warm enough to ensure that it would be serviceable when it came time to leave. The stay on the ice gave Weldy time to make a decision that was to change the face of Arctic transportation for years to come: he could extend light aircraft operations throughout the dark winter.

Duties carried out, pilot and doctor re-established visual contact with their guardian angel Dakota and bounced back into the air for the return to Resolute.

While I was having a drink with Weldy a short time after that trip, he mentioned almost accidentally that he had been out of town for a few days – up north. The story, brief as it was coming from the enigmatic Weldy, revealed his obvious pride in the accomplishment. It was a pride perhaps as much for the validation of his Phipps Special Super Cub under such severe winter conditions as for the humanitarian contribution to the Inuit of Grise.

~~~~~~

The Grise Fiord mercy flight was mentioned in the Lives Lived column in the Globe and Mail which I submitted following Weldy's death October 29, 1996. The obituary was followed by a letter to the editor dated December 27:

## WELDY PHIPPS

> The mid-winter flight to Grise Fiord by Weldy Phipps was but one of the Arctic rescue ventures by this remarkable Canadian aviator. (Lives Lived – Dec. 17).
>
> In 1959, I served as meteorologist and Canadian liaison officer on an American expedition to the ice shelf of Northern Ellesmere Island. Sponsored by the United States Air Force to determine if bombers could land on the shelf, the expedition produced little of scientific value and nothing of practical use. It suffered

from the American penchant for excess – too much technology, too many people. When our glaciologist, Paul Walker, suffered a stroke, the Americans had to send for Weldy.

We cleared some of the larger rocks off an impossibly small patch of land on Ward Hunt Island. With no navigational aids to guide him, Weldy bumped down on the oversized tires of his Phipps Special, picked up his passenger, and flew him to Thule. Sadly, Paul died later in California. The hill on Ward Hunt Island is named after this fine young American.

There are many unnamed peaks on Ellesmere Island. I hope that the Board of Geographical Names in the Northwest Territories will name one after Weldy. In winter dark and summer sun, in a land of beauty and solitude, the memory of this remarkable Canadian will remain green in the country he loved and served so well.

### Jim Lotz, Halifax

During his four-month assignment Jim Lotz camped at the edge of the ice shelf where he operated a meteorological station. Weldy Phipps had been contracted to provide some local air transportation for expedition glaciologist Paul Walker. He was guaranteed three hours flying a day at $40 an hour. Paul had flown over a great deal of Ward Hunt Island with Weldy and built cairns at the snouts of four glaciers. Mostly they had been visiting the glaciers to the west and east of the island camp making use of about 20 gas and food caches Weldy had strewn around the area. At Lake Hazen they set down and replenished the larder with Arctic char. In addition to scientific information Paul also received what might be the the most northerly mosquito bite in Canada.

Some time later, while Paul was on an overland trip from Ward Hunt Island, he suddenly had a pain in the base of his skull. Via radio, the doctor on the Coast Guard cutter Westwind said that it sounded like a pinched nerve. But Paul's right side was paralyzed and he had severe headaches. American authorities at Thule, Greenland, were radioed for a medical evacuation. But only Weldy Phipps was capable of making the tricky landing at the camp with his Phipps Special Super Cub.

Weldy left Resolute for the 600-plus mile emergency flight at 12.15 a.m. and arrived at 8 p.m. But, Jim Lotz added: "It might have been 8 a.m. We tended to lose the sense of time because the sun stayed up all day. I noted that he made 'very good time'. We cleared rocks from a stretch of open land on Ward Hunt Island – we did not think Weldy could land there. But he did. He took off at nine with Paul to fly him to Thule, another 650 kilometres over hostile terrain at less than 150 km/hr. Fifteen minutes after they left, the fog rolled in to obscure the entire area."

"Paul was flown from Thule to the United States." Jim Lotz said. "I went to see him in Pasadena. He was at home, paralyzed, where he later died. I suggested that the hill on Ward Hunt Island be named after him and this recommendation was accepted. He was a fine man, a good scientist and a true friend."

Weldy was a 'broad, friendly chap', Jim Lotz noted in his diary of those times. "We talked over tea and he gave us all the gen on what was happening in the High Arctic. I have another diary entry on Wednesday, July 8, 1959, noting that Weldy landed on the far side of Ward Hunt Island from our camp in thick fog. He was redirected by an expedition member and landed behind the camp in about 200 feet, turning sharply at the last second to miss an antenna."

A further diary entry: "I have a tremendous admiration for Phipps – he really pioneered the use of light planes in the Arctic. I wrote him a note, sending him authorization to draw gas from Hazen. God knows if this is within my power but it's too damn bad if it isn't. Weldy was someone you wanted to do your best for." An additional notes says: "Phipps has impressed everyone as a quiet guy, who snores."

~~~~~~

## Knee-jerk reaction...

The process of taking parts from one unserviceable aircraft to repair another is known in aviation circles as cannibalization, and it was one of Weldy's favourite solutions to technical problems. Spartan navigator Ray Lachance recalled the time at Resolute about 1958 or 1959 when aerial survey operations became bogged down by the unserviceability of two Mosquito aircraft. Each had a bad engine.

When Weldy landed his Super Cub at Resolute he saw the two aircraft and then met their crews idling at the bar. As former chief pilot of Spartan he asked, "Why are you guys sitting around when you should be at work?" The engine trouble was explained. "Well, why don't you take a good engine off one aircraft and replace the bad one on the other?" The crews thought he was joking.

"We can't do it," was the easy reply.

"Sure you can. I'll give you a hand," Weldy said. With nothing to lose the Spartan crews agreed to give it a try. They set to work at 5 p.m. and had the one aircraft ready to go at 7 a.m.

"There was nothing in it for Weldy," Ray Lachance said. "He was a partner in Bradley Air Services at the time but he couldn't resist the challenge to keep one of his old aircraft in action."

~~~~~~~

## The hunt for minerals...

The acid test for the Phipps Special tires came in 1958 in a contract to provide one aircraft to the Geological Survey of Canada. Following a summer of explorations Dr. Ray Thorsteinsson and Dr. E.T. "Tim" Tozer composed the following interim report on the success of the Phipps aircraft in their work:

## Geological investigation of Melville, Prince Patrick, Eglinton, Brock, Borden and Mackenzie King Islands, Canadian Arctic Archipelago, 1958.

During the summer of 1958, a field party of the Geological Survey of Canada, conducted a reconnaissance survey in the northwestern part of the Canadian Arctic Archipelago using a Piper Super Cub PA18A aircraft. The survey was conducted as an experiment as this was the first time that a fixed wing aircraft equipped with wheels designed for landing on unprepared terrain has been employed in geological mapping in Canada.

The survey was necessarily small. It consisted of two geologists, R. Thorsteinsson and E.T. Tozer, jointly in charge, and W.W. Phipps, a licensed pilot-engineer. The aircraft was provided under contract by Bradley Air Services, of which company Phipps is Vice-President. This company has its headquarters at Carp, Ontario, about 20 miles west of Ottawa.

The main base of operation for the Piper Cub survey was at Tingmisut Lake, at the south end of Sabine Peninsula, Melville Island, where some 2,500 gallons of avgas and 600-man-day food rations had been cached in 1955. This cache was established by DC-3 aircraft, equipped with ski-wheels, in preparation for the helicopter-borne geological survey known as "Operation Franklin" (See The Polar Record, vol. 8, No. 53, p. 157).

In the planning of Operation Franklin the cache on Sabine Peninsula had been planned as the last of a series of bases for occupation as the operation proceeded counter-clockwise through the Queen

Elizabeth Islands. However owing to adverse weather conditions the site of the cache was not occupied for any length of time and most of the fuel and food was available for the Piper Cub operation.

In order to permit landings on unprepared terrain, Phipps equipped the aircraft with large balloon tires *(25" x 11" x 4" and bearing 5 lbs pressure) which are standard equipment on the Stinson SR5. With slight modifications the hubs of the super Cub wheels were readily adapted to the Stinson's tires. The Piper Super Cub is powered by a 150 h.p. Lycoming engine. The increased drag of the large tires reduced the normal cruising speed of about 115 m.p.h. to about 95 m.p.h.. Fuel consumption is about 6.5 gallons per hour, the fuel tanks carry 30 gallons which permit a range of about 400 miles. The take-off run is about 200 feet and the landing roll is somewhat greater. Both, of course, depend upon wind speed. In Canada the Super Cub is licensed to carry a disposable load of 700 lbs.

~~~~~~

* [Note: The 25-inch tires, the first balloon tires used, came from a Lockheed 10. Later tires were 35 inches in diameter. The hubs were from a Stinson SR5 Reliant.]

~~~~~~

**The main objectives of the 1958 Survey were:**

    (a) to delimit and describe the bedrock formations and

    (b) to produce a geological map on the scale of 8 miles to the inch.

A critical study of air photographs formed an important part of the preparations prior to the field studies. From air photographs and previously acquired geological information, photogeologic maps were prepared as working hypotheses and from these maps critical areas were selected for ground investigations.

The aircraft left Carp Airport on June 15. In order to cover the long distances between refuelling points Phipps carried an additional 45 gallons of avgas in a drum strapped behind the pilot's seat. Phipps designed a pumping system which enabled him, while airborne, to fill the wing tanks from this drum, plus full wing tanks, permitted a range of about 1,000 miles. Phipps' route lay through Weenusk, Ontario; Churchill, Manitoba; Baker Lake, N.W.T.

On June 19 Thorsteinsson and Tozer were flown by the RCAF from Churchill to Resolute Bay, on Cornwallis Island. On the same day Phipps arrived at Resolute. On his way north he had spent two days based at the camp of T.H. Manning and Mr. and Mrs. Andrew Macpherson, biologists engaged in a survey of Prince of Wales Island for the National Museum of Canada.

On June 20 the party started the airlift to Melville Island. Five flights between Resolute and Tingmisut Lake were required to transport the geologists, their equipment and instruments to the base camp. Geological investigations started on June 25 when a temporary camp was established on the north side of Raglan Range, northwestern Melville Island.

On July 11 the party moved to Mackenzie King Island and set up a temporary camp near the Leffingwell Crags. Ten days were devoted to geological study of Brock, Borden and Mackenzie King Island. These landings represent the first visit to Brock, Borden and Mackenzie King Islands since their discovery by the Canadian Arctic Expeditions (1913-1918) led by Vilhaljmur Stefansson.

Between July 21-26 the party made its headquarters at the Mould Bay Weather Station, Prince Patrick Island. From Mould Bay flights were made to points on Prince Patrick and Eglinton Islands. On July 26 the party returned to Melville Island and a temporary camp was established on the north side of Ibbett Bay. Here the party remained for a period of four days after which they returned to the base camp at Tingmisut Lake. After completing the geological study of Melville Island on September 1 the party started the return airlift to Resolute. Adverse weather delayed this airlift and final evacuation of the base camp was not completed until September 10.

During field work radio communication was maintained between the aircraft and the weather stations at Resolute and Mould Bay. Daily flight plans were filed. A 2-watt radio at the main base camp made intermittent contact with the aircraft and the weather stations. Emergency and survival equipment was carried on all flights. Owing to the great distances over which the survey ranged it was necessary to set out several small caches of avgas. Such caches were deposited on Mackenzie King Island and on western Melville Island at Ibbett Bay and Cape Fisher. When occupied in depositing caches of fuel the aircraft was capable of carrying four 10-gallon drums of Avgas.

A total of 300 hours were flown during the course of the field season. This figure includes 60 hours used in flying up to Resolute and back to Carp. Approximately 450 landings were made of which about 400 were on unprepared landing surfaces. The aircraft landed (on unprepared) terrain in 249 different places. One hundred and sixty of these were on Melville Island, 29 on Mackenzie King Island, 27 on Prince Patrick island, 15 on Borden Island, 10 on Eglinton Island, 4 on Brock Island, 3 on Cameron Island and 1 on Emerald Island.

In the course of field work the aircraft served to transport the geologists to critical outcrops and it also served as a vantage point from which formational contacts and other structural features could be recognized and located accurately on air photographs and maps. For each geologist the field work involved an alternation of one day's flying and one day in camp compiling field data and planning the following day's air traverse.

The islands investigated and mapped comprise an area of 23,800 square miles. Some portions of this area had been studied by Tozer in 1954 and 1955, travelling by dog team and on foot, but most of the area had not been previously studied by the Geological Survey.

The performance of the aircraft was satisfactory in every respect and the investigation proceeded without mishap. On six occasions during the survey the aircraft became mired in mud on landing. This occurred as the aircraft came to a halt and the weight of the aircraft settled entirely upon the wheels. In general, miring occurred only on surfaces which would not support a man on foot. The depth to which the aircraft sank on these occasions seldom exceeded the height of the hubs.

The depth of sinking was determined by the depth of the frost table, which in these latitudes and in the soil encountered generally lies within a few inches of the surface. Extrication from such mud was achieved by one man lifting on the wing strut while the other constructed a support of rocks under the tires. In this manner the aircraft was gradually worked on to firm ground.

While engaged in geological work the geologist would choose an area where he wished to land. On every occasion, with one exception, Phipps was able to land within ten minutes walking distance of the chosen locality. The one exception was caused by unusually high winds rather than unsuitable terrain.

Three characteristics of the Piper Super Cub that contributed to its efficiency in the 1958 Survey are as follows:

1) the buoyancy of the large tires on soft terrain;

2) the shock absorbing properties of the undercarriage, which was dependent upon the soft tires and the conventional shock absorbers;

3) the outstanding, low speed handling characteristics that permit a safe approach and landing at an indicated speed of 35 to 40 miles per hour.

Absolutely calm conditions are rare in the area studied, consequently most landings were made with a head wind that reduced the ground speed on landing to less than 35 miles per hour.

~~~~~~

**Some results of northern operations:**

- The 1958 expedition produced a 50-fold increase in efficiency in the geological survey;

- The two geologists covered 30,000 square miles in 300 hours of flying. (by canoe and dog team only 400-600 square miles could be investigated, mainly only in coastal areas.

- Cost (1958) $12,000 compares with Franklin Expedition which used two helicopters and 26 geologists to cover 60,000 square miles at a cost of $300,000 ($200,000 for the helicopters). Further such investigations abandoned because of cost.

- In 1959 five Super Cubs covered 100,000 square miles on Banks and Victoria Islands under contract to the GSC.

- In 1959 Phipps carried out a mercy mission for the USAF, airlifting an ailing man from Ward Hunt Island to Alert in a 700-mile dash. American aircraft could not land at the point and their helicopters could not be staged to the area in time.

- In 1960 only the low cost of the Cubs enabled the Jacobsen- McGill Expedition to collect vital scientific information around Axel Heiberg Island.

- In 1961 the Department of Northern Affairs was able to carry out the first comprehensive survey of migratory birds only by using the Cubs.

~~~~~~

## Chapter 5
### The rebirth of Atlas...

It was New Year's Eve 1961. And as 1962 came about, the temperature dropped to its seasonal low, an ideal time for the furnace at the Carp headquarters of Bradley Air Services to go on the blink. It wasn't a very serious malfunction, just a balky thermostat, but frozen water pipes would turn the situation into a crisis. Since he was in the process of adjusting from active flying in the north to duties back at Carp, Weldy volunteered to keep the furnace running. It was a humdrum job, just waiting for the thermometer to reach uncomfortable levels before he had to restart the burner manually.

Having been away all summer he had been out of touch with the business end of the company. Now he had time on his hands, enough to ferret through some accessible documents to get a feel for the health of the organization of which he was vice-president, chief of operations and one-third partner. What he found, he said, "didn't look right" to him. It appeared that his partner, Russ Bradley, had been part of a scam that had bilked the federal government by selling the Department of Transport aircraft parts it already owned.

To recheck his suspicions he enlisted the help of Carl "Soggy" Norton, a detective with the Ottawa police department. It wasn't a routine investigation and not one within the jurisdiction of his Ottawa duties. But friendships forged in wartime prison camps transcended the rules. Soggy undertook a private investigation of the things that didn't look right. His probing took him to Belleville and Montreal, where it was discovered that two of the aircraft firms recorded in the transactions were fakes.

Weldy collected the questionable invoices and relevant documents and took them to an Ottawa lawyer. These papers were subsequently deposited with an auditing firm for review. The review resulted in a declaration by the auditors to lawyers representing both Phipps and Bradley, that the government had indeed been defrauded. Their report concluded that the government had paid for repair work never done and paid for aircraft parts that had been sold fraudulently to the DOT with the co-operation of "a senior DOT official".

At a subsequent meeting in the auditors' offices, according to a report in the Ottawa Citizen, Russ Bradley admitted that the allegations made in the report were accurate. It was agreed by all concerned that the matter would have to be disclosed to authorities.

The matter was presented to the Hon. Leon Balcer, Minister of Transport. Bradley was charged with defrauding the government to the extent of $54,565 for work supposedly done on Department of Transport aircraft between 1959 and 1961, and which was never actually done; and defrauding the government in the amount of $25,000 by selling to the DOT aircraft magnetos that were already government property.

The case gained momentum with the simultaneous indictment of Henry John Cashell, former head of the DOT depot at Ottawa Airport. Cashell pleaded guilty to the same charges. In court he outlined the system he had set up, which included two fictitious companies. One with a postal box address in Trenton and bank account in Belleville was called J.B. Scarrow Sales and Service. (Cashell used the alias Scarrow.) That company, he said, sold magnetos to Bradley for $800 each. The units had already been purchased from a Toronto company and paid for by the DOT. He further testified that Bradley paid Scarrow up front and later sold the units to the department at a higher price. Another phony firm had been set up by Cashell with a Montreal address. He said a bank account under the name Jackson was opened in Gananoque, Ontario.

Cashell testified under oath that Bradley had been a part of the scam, always insisting that he pay in cash rather than by cheque in his transactions. Prominent Toronto lawyer Arthur Martin countered with the presentation of cancelled cheques signed by Russ Bradley. Cashell had also testified that when he was paid by cheque he had paid a kickback to Bradley in cash. With Cashell's credibility in doubt, the jury acquitted Russ Bradley. Cashell had already been sentenced to three years in penitentiary. The incident led to the dissolution of the Bradley-Phipps partnership and set Weldy off on his own.

(Russ Bradley died in the fiery crash of his Cadillac just west of Ottawa in late 1970.)

The hastily-formed Phipps company needed a name, and Weldy had no trouble coming up with his first choice – Atlas. It was a sentimental attachment to his first job in commercial aviation. The name, however, was still alive and in the custody of Angus Morrison, original president of Atlas Aviation Ltd., which had ceased operations in the fifties. Could Weldy buy the name? "No," said Mr. Morrison, by then president of the Air Transport Association of Canada, "You can have it with my compliments." The sentimentality for the Good Old Days ran strongly in both directions.

The new Atlas began rather humbly on Doc Demerah's farm near Richmond, Ontario. The first company office was a large Pontiac station wagon with a radio-phone in it. The Phipps home on Thorson Avenue in Ottawa became the temporary repository of aircraft parts. At the farm, some aircraft had to be put in flying condition and sold. The first revenue-producing aircraft was a Beaver, used for a while to haul seal pelts off the Gulf of St. Lawrence ice. But Weldy disliked the job because of the mess it made of the Beaver. His eye was on the High Arctic where he knew the lay of the land and his ability to survive there. His experience on the Grise Fiord mercy flight had convinced him that year-round operations were viable. He moved to Resolute on his own.

If the name Resolute referred to the determination of any man, the term could be fittingly applied to Weldy Phipps. For it took a resolute flyer to choose such a wild climate in hostile terrain as a territory in which to operate. Temperatures range from a low of minus 40C in January to highs in July of one to seven degrees. There is little snowfall, however frequent gale force winds create heavy drifting, often burying exposed aircraft up to the wing roots. There is little atmospheric comfort to anyone flying into or out of Resolute Bay throughout most of the year. At least 42 aircraft had crashed on Cornwallis Island alone when Atlas took up residence. Others are scattered around nearby islands.

From Resolute Weldy contacted Fran with a rather tentative suggestion that the family join him. It was not an idea that Fran pondered for very long. She sold their comfortable home and moved the family to a ragtag collection of portable buildings on the Resolute airport.

~~~~~~

## Y'all cain't do that in Thule!

Learning the Arctic ropes was a great experience for the Phipps family. The 'bright lights' might be seen on a trip to Iqaluit or to Thule, Greenland. But as Fran and Weldy learned, Thule had a culture with a difference.

A fall from the top of a fuel storage tank at Resolute resulted in an urgent medevac of the injured man to the nearest hospital, at Thule. There was no available nurse to accompany the flight and so Fran was pressed into service. She felt unequipped to perform duties for which she had no training but was assured that she only had to keep the patient comfortable.

Social life in the neighbourhood of Resolute was limited and so the prospect of visiting other people only 2000 km away was very attractive to Fran. Accommodation at the U.S. military outpost, however, posed a problem. She learned of the difference when she made a frantic request, upon landing, for a bathroom. An armed soldier escorted her to the men's barracks and pointed her into the facilities. But instead of a normal bathroom she faced a long line of naked toilet bowls – no cubicles, no doors. But the urgency of the situation exceeded modesty and after all there was an armed soldier outside the room standing guard.

Surprise No. 2: Male and female guests on the American Air Force base, married or not, had to be housed overnight in separate buildings. It seems that the men of the base, so far from home, tended to suffer female deprivation at the critical level. That made Thule – with a ratio of 5,989 men to 11 women at the time – a hardship posting for randy servicemen who might have found a war zone more to their liking.

Weldy was assigned a room in the men's quarters and Fran in the limited women's. But Fran was not impressed, and decided to find a way to avoid the segregation rule. She bundled her parka around her head and tried to look as much like a man as possible. Walking with Weldy she modified her usual graceful gait to a trudging stomp. She entered the men's quarters undetected, and once in the room, began making preparations for bed. But a note was slipped under the door by another passenger on the flight. It said that the Military Police were searching for "a woman named Fran." Quickly, she tried to scramble under the bed. But the bed was too low. She spent the rest of the night in more of a stupor than sleep.

When the early morning came, Fran was bleary-eyed. She rearranged her parka and sneaked out of the building gradually converting her masculine saunter to a dainty stroll with wiggle as she approached the women's quarters. There she was able to take her time to freshen up. She was still attending to her ablutions when Weldy was ready for the return flight to Resolute.

Surprise No. 3: Y'all cain't even phone the women's quarters. This offered little challenge to Weldy who simply went up, opened the door, and shouted "FRAN!" That's all it took to ignite a scene worthy of the Keystone Kops as MPs swooped in to arrest Weldy. How could he know that the building was hot wired to the Military Police? The interrogation that followed met with a culture clash between the frontiersman and the down south soldier. The MPs finally ran out of logical argument and found it simpler to let the intruders go. Whiskey Whiskey Papa lifted off and turned westward to Resolute.

~~~~~~~

## A Labour Not Quite of Love

Medical evacuations in the Arctic dark period were not the most attractive charters in Atlas's scheme. But one such trip came up in midwinter. The nurse at Arctic Bay urgently wanted a pregnant Inuit woman airlifted to Frobisher Bay. The weather: minus 20C in blowing snow. Weldy wanted to postpone the trip, but the nurse felt it was too risky.

Against his better judgment, he took off in a Twin Otter and headed to Arctic Bay, some 200 miles away. The weather was no better there. He called ahead to have flare pots set out. After a safe landing he picked up the Inuit woman who had by then been in labour for 36 hours. They flew to Frobisher where she was delivered to the medical centre. Two weeks later she delivered her baby. The labour emergency turned out to be severe constipation.

~~~~~~

## The art of staying alive...

Weldy's lectures to pilots stressed the importance of reading landing areas carefully. Many sites looked quite suitable from the air, but proved much less so at touch-down. Of particular concern were river beds which could look flat, but end up miring the aircraft, if not causing greater havoc. He had acquired some first hand experience with such a site at Fort Conger when he pranged a Super Cub.

Among his other rules were to land on high ground where it looked safest. That safe appearance could be deceiving on river bottoms, outwashes and banks unless a firm base of gravel could be identified from the air. Since the Arctic has a stiff breeze, if not a strong wind, blowing most of the time it was to be taken advantage of on landing. But just in case, try to land up hill (but not directly into a cliff). Plan to have the aircraft come to a stop on the smoothest area.

Landing on the sea ice, Weldy would reverse the props on the Twin Otter just before touchdown, effecting a very short roll.

A Bristol Freighter awaits its eventualy plunge to the ocean bottom at Pangnirtung. Weldy's dive in an attempt to salvage it was too late.
—Phipps collection

On takeoff he was meticulous about loading. Flying a single Otter on one occasion, Weldy supervised its loading carefully to ensure a safe weight and balance. Then he went for a final cup of coffee. In his brief absence local workers continued to load the aircraft. Weldy discovered the imbalance in the air and had to fight the controls to remain airborne. Twin Otters normally took off in 750 feet and landed in 500 feet or less.

Serviceability of the aircraft was paramount. That meant avoiding one of the main problems – contaminated fuel. At times it became difficult to tell just what was in a barrel that might have been around for years. What was thought to be aviation gas could turn out to be naphtha or diesel fuel. And avgas that had been warmed and rained upon for a couple of seasons could contain water ingested when the barrel cooled.

~~~~~~

### Getting around the Arctic...

Navigation in the High Arctic demanded keen skills in the era of limited navaids and inadequate or non-existent charts. The variation at Resolute is 80 degrees west, making a magnetic compass useless. The Atlas aircraft carried an astro compass on which the pilot set the local hour angle and approximate latitude then aimed the instrument's rifle sights at the sun. The true heading could then be read and the directional gyro reset. During dark winter months star sights might be taken with a mariner's sextant. The Twin Otter used a drift meter, a periscope looking backward to detect drift. A radar altimeter kept accurate track of the height above ground.

~~~~~~

## Navigation by Finger

The Inuit are renowned for their ability to navigate their way across featureless lands. Along the way they often set up inukshuks to indicate where they have been or where they have gone. Navigation across the same lands by air, when snow covers everything, posed a daunting problem for Weldy in 1954. When he was challenged to perform survey work he decided to take a native with him to guide the route to Gary Lake. Navigation consisted of the native pointing his finger. Weldy followed direction on his understanding that these people were wizards at finding their way. After a lengthy flight he began to worry that they would never find Gary Lake. Then the enigmatic Eskimo pointed straight down. Weldy interpreted the signal to mean that they had arrived. He looked down and there was Gary Lake camouflaged in snow.

~~~~~~

## Blame it on the compass...

On one contract Weldy had to locate certain targets to very exacting tolerances. To ensure accuracy he took along a navigator. As the mission progressed it became evident that they were off course. The navigator finally had to admit his error.

Weldy looked down and spotted an Inuit settlement. He landed the aircraft and said to the navigator, "Go over to the village and ask them where we are." With the navigator trudging to the community, Weldy took off and flew away. He returned to pick up the navigator – a week later.

That was one of those tales that was hard to confirm. But in an attempt to cross-reference the incident an informant said, "I can't vouch for that one but I believe it." He went on to tell about an aircraft that landed on the ice at Pangnirtung. The ice had opened up. The Bristol Freighter sank. Now it was a salvageable aircraft and Weldy was anxious to acquire an aircraft at a bargain price.

He hired an expert Arctic diver who flew to Pangnirtung for the salvage job. But when the diver saw the ice and current, he declined to make the dive. Exasperated, Weldy said, "Well give me your suit, then." He dove on the aircraft himself and determined that the shifting ice had further damaged the Freighter beyond economical repair. He handed the diver his equipment back. The expert diver demanded his fee.

"For what?" Weldy asked. "All you did was watch."

~~~~~~

## Expo 67 film…

In preparation for Expo '67, a party of filmmakers was shooting in the High Arctic for the spectacular multi-screen production for the Man and the Polar Regions pavilion. This work, which had included shooting around both poles, chose Atlas Aviation as their carrier.

Since it was to be a circular theatre in which the audience rotated while viewing two adjoining films as one, the crew required two cameras shooting at diverging angles simultaneously. For an accomplished aircraft redesigner, Weldy had no trouble mounting the two 35mm cameras on the nose of a Piper Apache. While preparations were made, and they waited for the fog to lift, the crew filmed some location shots around Resolute. In the radio room, with camera and sound recorders rolling, a mayday distress call came in. A pilot, low on fuel couldn't land because of the fog. In his first attempt the distraught pilot missed the runway and had to go around. Somebody ran to get Weldy.

"Tell him to swing out over Barrow Strait and turn due north," Weldy said. "I'm going to build him a fire he can see for 100 miles." Weldy was recycling a technique he knew from Bomber Command days. Returning bombers unable to land at

An early collection of Bradley aircraft in the north show the Phipps Special tires that made operation on rought terrain possible.

—Phipps collection

their own bases because of fog could use runways equipped with FIDO (Fog Investigation Dispersal Operation). The long runway in England was bordered by perforated pipes carrying gasoline. Jets of gas were ignited, creating a wave of heat that diminished the fog well enough to permit a landing.

Weldy had used a primitive version of the technique before at Resolute. Faced with the latest emergency, he ran off and found an RCAF truck carrying seven drums of gasoline. At the end of the runway, he dumped the gas and tried to light it. But the fog rendered all available matches useless. Most of the Resolute airport population stood at the runway's edge. All was quiet. Then suddenly out of nowhere the Beechcraft roared inches over their heads, sending all the spectators sprawling on the ground. The pilot had spotted a wispy hole in the fog and made a desperate but successful landing. Most of the excitement was captured on film, much to the satisfaction of the film makers. It was their only real drama for the production.

Twenty minutes after the plane landed, the ashen-faced pilot, an American who was trying to show his son the Arctic, came to thank Weldy for his advice and the effort to light the fire. "If there is anything at all I can do for you..."

"Nothing for me," Weldy said, "But you owe the air force seven drums of high test." The next morning when activity resumed around Resolute, the American aircraft had already departed.

Eventually the fog lifted. The movie crew boarded the laden Apache for the shoot. One of Weldy's pilots, John Strickland, flew them through canyons and around peaks as the nose-mounted cameras captured the breathtaking scenery. Because of the heavy drag imposed by the cameras, the Apache struggled to stay airborne. Full power was required to maintain level flight. And then it happened. Out went an engine. The aircraft began to sink.

"Jettison everything," the pilot shouted. And the crew grabbed everything they could get their hands on, including valuable reels of exposed film, tape recorders, and ancillary equipment. One technician, an Austrian, even threw his passport out by mistake. The aircraft stopped sinking and limped anxiously to a safe landing. The engine restarted easily on the ground and offered no further trouble.

The connection between one distressed flight and the other is one the film's producers, Robert Kerr and Graeme Ferguson, would rather forget. Their windfall Arctic drama is still out there among the muskoxen and polar bears. Only a small snippet of the emergency landing by the distressed plane remained. It showed the beaming faces of the Resolute people when the aircraft taxied back out of the fog, Fran Phipps embracing the wrung-out pilot as he slid down from the cockpit. The Arctic folks are like that.

The otherwise successful film, Man and the Polar Regions, which used an innovative rotating theatre, served as a springboard for Kerr and Ferguson to found the world-famous IMAX Entertainment empire.

~~~~~~

## Mr. Fix-it, Arctic variety...

Weldy had a reputation even in Ottawa's Hilson Avenue Public School of endless tinkering with anything that had two or more parts. It was a curiosity that made him a master technician. But in his self-styled technical grounding, there were many Ottawa Valley barnyard solutions to problems.

There was the time when he was flying the principal of McGill University along with other members of an Arctic scientific expedition. In flight, the windscreen of the aircraft was suddenly splattered with oil. The passengers became alarmed.

Weldy, in typical cool fashion, landed the Otter. "Has anybody got a piece of cardboard?" he asked his passengers. A piece of a cornflakes box was found. Weldy loosened the plane's prop, calculated the size of gasket required to stem the escape of oil from the pitch-changing mechanism, cut one out, installed it, replaced the prop, wiped off the windscreen, and continued the flight.

~~~~~~

## If I had a hammer...

Another highly technical operation was performed on a brand new Twin Otter that Dick deBlicquy had landed at Tanquary Fiord. He had set it down on a wide swath of the Air Force Glacier to pick up members of a British and Canadian mountain survey party. The Brits were from a Royal Air Force "hardship" duty detachment and the Canadians from the Defence Research Board scientific base at Tanquary. But when it came time to leave, Dick could not get off the glacier because the starboard engine refused to operate above idle. This forced Dick and his would-be passengers to hobble most of the 40 miles back to base.

Dick called Weldy on the base radio. "Get a hammer," Weldy said, "and tap the fuel pump." The RAF members were aghast. But before this could be accomplished, Weldy diverted from transporting tranquilized caribou between islands, and landed beside the ailing aircraft. He warmed the fuel pump, then tapped it gently a few times. The engine came to life and the aircraft was able to depart.

Weldy's comment: "Every aircraft should carry a hammer."

~~~~~~

## The tailshovel...

The unusual appendage scraping along the tarmac drew the official attention of a transport department inspector at Churchill airport. But Weldy couldn't quite understand the fuss. He had broken the tailwheel off a Super Cub somewhere in the High Arctic. Replacement parts were at least a thousand miles away and so it made sense to improvise. Weldy found a shovel, cut the handle down and fixed the blade at a trailing attitude where the wheel was supposed to be, and taxied on. In another tail-twisting mishap far from supplies he repaired the vertical stabilizer on a Super Cub with two ski poles. And yet another field repair was made with a sheet of plywood.

~~~~~~

## Getting up steam...

Through much of his working life Weldy Phipps kept hearing the same thing: "You can't do that..." His response was always the same: "Let's try..."

Dick deBlicquy made such a comment at Resolute Bay in the early sixties. Atlas Aviation had a lot of empty barrels that had been shipped north by barge. Empty, they were still worth about $25 each and Atlas could not afford to abandon them. To get the value out of them, however, they had to be loaded onto a barge to be taken south. Shipping regulations stipulated that the barrels must be steam-cleaned as a safety precaution. That posed a problem. The only steam generator belonged to the Department of Transport and they would not lend it out. They still had their own barrels to steam.

"I think we should just fill a barrel with water," Weldy said, "build a fire under it and harness the steam." Enter Dick the naysayer, who was learning after a dozen years working with

Weldy, that it was pointless to deny the possibilities. They tried it, and it worked like a charm. All the barrels got steam-cleaned with the homemade rig and shipped south for the rebate.

~~~~~~

## The Turbo Beech...

One day Weldy approached Doug Irving at Ottawa airport with some drawings for an air scoop. They related to a turbocharger modification he was planning for a Beech 18 aircraft he wanted to use for high-level photography. Doug Irving, president of Alexander Metals, was not only a superb metals craftsman but a private pilot as well. He was, at the time, president of the Ottawa Flying Club. He had no trouble following Weldy's plans for an air scoop and plumbing that would feed ram air to the turbocharger for the two engines on the Beech.

"Weldy got a turbo charger either off a P-38 or a B-17," said Bill Doherty, "they are basically the same thing. He brought pipes back from each engine and mounted them in the center of the belly under the cockpit." The scoop was fitted, and Weldy finished the installation of the turbocharger. He could hardly wait to test fly the boosted aircraft at the target 20,000 feet.

It was some time before Doug saw Weldy again. But next time he saw him he asked: "How did the turbocharger work?"

"Like a charm," Weldy replied. "I got almost to 20,000 feet before I realized that the wings were just not going to fly any more. I got into the worst spin in my life and only recovered when I got close to the ground."

After that, the Beech spent a lot of time parked at the Ottawa airport, no longer a high-altitude mapping aircraft. The modification proved to be one of Weldy's rare miscalculations. But the aircraft provided a welcome nesting place for the airport's sparrows. Bill Doherty concluded: "There wasn't many things he didn't try."

~~~~~~

## A tern for the worst...

His colleagues from wartime and the most difficult of civilian flying operations often commented that Weldy was a man without fear. But Weldy counted as his worst moment of fear the time he was refuelling an aircraft, standing on one wheel. The whole Arctic seemed silent at the time. Suddenly an ear-piercing scream came out of nowhere and something big brushed his head. He flew one way – the gas can, chamois fuel filter and funnel went the other way. He had been dive bombed by an Arctic tern which resented his proximity to its nest.

~~~~~~

## Mark Twainesque weather...

Mark Twain had a thing about weather: everybody complains about it, but nobody ever does anything about it. In 1876 he said in a speech: "There is a sumptuous variety about the New England weather that compels the stranger's admiration and regret. The weather is always doing something there; always attending strictly to business; always getting up new designs and trying them on the people to see how they will go. But it gets through more business in spring than in any other season. In spring I have counted one hundred and thirty-six different kinds of weather inside of four-and twenty hours."

Too bad old Mark couldn't have visited the High Arctic. He might have added a variety or two. The muse, however, was there in the pre-satellite days. Somewhere between periods of intense activity and hours of boredom imposed by fickle Arctic weather, the muse was stirred. This literary offering is believed to have been penned by Barry Ralph who was dispatcher for Atlas Aviation in Resolute:

## The No-name Poem

Drinking beer and smoking fags,
Pilots like to chew the rags.
They sit in circles in the bar,
Telling tales about the forecaster.

How he saves their worthless skins;
How he bought them seas of gins;
How he toils both night and day
For half their huge amount of pay.

In dulcet tones they bless this man,
For his uplifted benedictory hand,
But never do they pay him due.
They are a cagey lot, this pilot crew.

Flying for the strangest reason,

 They fly right through the blessed season.
When good men all lie sound asleep,
Pilots search for a forecast sheet.

With mincing step they stagger in,
With bleeding eyes they glare at him.
Wishing to heaven they hadn't drunk,
Or told those stories so full of bunk.

They try to listen to the cultured tones.
But half asleep with spavined bones,
These simple men are lost in awe
At the flowing wisdom of the forecaster.

This superior being called out of bed
Explains the weather to these great fat heads.
They never warn him of their needs,
As is required by all but Swedes.

They ask for winds, but little note
The direction, speed or the other dope,
And when they miss their destination,
They all intone, "t'was misdirection".

With patience much renowned,
With a true tongue so seldom found,
The Met man builds his reputation
To offset these ululations.

That is why all him revere,
That is why these pilots rear
Upright when he approaches them
To give them something within their little ken.

Ah, well! Noblesse oblige,
T'is likely to be a lengthy siege.
And only true men can maintain
The happy medium with those insane.

Arctic weather presented a big challenge to both air and ground crews – 100-mph winds, 24-hour nights and minus-60-degree temperatures. There is fog: radiation, advection, up-slope, frost smoke, ice-crystal haze, ice-crystal fog and blowing snow. Ice crystals glint like cut glass if the sun is at the right angle; from the air they can cut visibility to five miles or less. And just to add insult to injury, aircraft exhaust can cause increased ice-crystal fog.

In the spring and fall – otherwise the best times to fly – icing is a hazard. The ice usually shows up in low-level stratus (it's rare that an Arctic pilot isn't flying in the clear at 5000 feet but occasionally it will be solid right on up to 10000. The Otters are particularly prone to icing problems: they pick up ice, as they

Twin Otters, a single Otter and a Beaver in front of the Atlas hangar on Phipps Field at Resolute Bay.

climb through layers of fog. As the aircraft pick up too much ice, not surprisingly, they sink back down again.

Wind is no treat either. When there's a crosswind at Resolute it's a beauty – 70 degrees to the runway, with gusts as high as 85 mph. A steady wind of 40 mph is ordinary. Weldy tells of landing his Twin Super Beech in a 70-degree, 70-mph crosswind. Until then it was the only time they had all the fire engines and crash trucks out at the same time at Resolute. Fran was sitting in the right seat, and Weldy had to ask her to lean way back to give him a view through her side window as they crabbed towards the runway.

Talking about it years later, Fran said that she wasn't particularly scared by the hairy landing. "I knew that if only one pilot in the world could handle the situation it would be Weldy. But," she added, "if we were going to be killed, at least it would be sudden."

That particular aircraft was the least favoured of the Atlas fleet. It required many hours of preparation for every hour of flight. In particular, it did not function well in winter. The frigid Arctic temperatures would seize the control cables. Doses of alcohol were required before takeoff to ensure that the controls would move the elevator, rudder and ailerons.

On another problematic approach to Resolute, Weldy had throttled back in preparation for landing. But the Super Beech suddenly entered a much colder zone, and the throttle control froze at the approach setting. There was not enough power to overshoot or continue flying at all. Weldy's only option was to pull back the mixture control to full lean and shut both engines down. He landed "dead stick".

The Arctic wind can change the summer environment in minutes as it turns and comes ashore over ice. A pilot faces the dilemma of flying in cloud and risking icing or being forced above cloud and losing visual contact with the ground. One recommended technique when trapped on top was to look for a land form that might be made recognizeable by the wind ripping through a gap. Once the feature is identified, the pilot can let down through the cloud to a blind landing on the known level spot.

Getting an aircraft down onto safe land makes the choice of options critical. On one occasion Weldy was unsure of the snow conditions on a mountain peak. He landed on a small shale outcrop sticking up through the snow. He then walked down to test the snow. It would have supported his aircraft.

Landing in unusual places was often dictated by convenience. At the Eureka weather station, where the airstrip is three miles from the station proper, Weldy would land the Cub behind the garage. And there was the time at the end of a northern contract when he agreed to fly the expedition leader home from Ottawa to Hudson, Que. Weldy performed one of his famous barnstorming sideslips right into George Jacobsen's back yard. Mrs. Jacobsen looked up suddenly to see a Super Cub on her lawn and nearly dropped the soup. There is an unconfirmed story that in a high wind one time he actually flew a Cub to a landing right inside a hangar.

Such landings did not always sit well with the passengers. When then-Prime Minister Pierre Trudeau was being flown to Grise Fiord he rode in the co-pilot's seat of the Twin Otter. A one-time private pilot, Mr. Trudeau grew increasingly apprehensive as the aircraft approached Grise directly towards the face of a cliff. By the time the aircraft landed, Trudeau had turned a couple of shades of white. When they rolled to a stop he caught his breath and said to Weldy that he wouldn't even try to park a Volkswagen there.

A freak downburst in near zero weather dashed this Super Cub onto the rock surface. Weldy suffered a serious face cut but his passenger was more seriously injured.

—Phipps collection

Many aircraft encountering High Arctic weather did not survive the experience. In the spring of 1970, an F27 with 13 passengers aboard, low on fuel, became trapped in a slashing snowstorm as it neared Resolute. Atlas employees could hear the pilot on the radio and there was no doubt that he was in deep trouble. Then the radio transmissions went silent. After ten minutes Weldy said, "He's down." With that he took off in a single-engine Otter.

Flying 200 feet above the rugged hills of Cornwallis Island, Weldy somehow spotted the tail section of the downed aircraft. He made six passes before he found a spot where he could set the Otter down. He ran to the wreck to find passengers and crew trying to walk to Resolute – in the wrong direction, a fatal mistake in that unforgiving land. All were dazed, two of them badly injured. He flew the two most badly injured to Resolute and then returned for the rest in the Twin Otter.

~~~~~

### In pursuit of the Pole...

In 1970 Weldy flew a Florida businessman to the North Pole. Word of the passenger's unusual trip spread rapidly. Very shortly after that flight, Atlas Aviation received more than 100 inquiries, mainly from the United States, for similar trips. That prodded Weldy's entrepreneurial spirit, and he began mapping out a strategy to fly more tourists to the geographical pole. He proposed a route from Resolute to Lake Hazen, where tourists would stay in a Quonset hut. A landing at the Pole required clear weather but that could never be guaranteed. The cost for a four or five-day round trip from Montreal or Toronto to the Pole at that time was set at $2,500.

Dick deBlicquy summed up the interest in tourism: "I've been to the Pole and there's not much there but a lot of ice and snow. Still, I can say I've been there, and I guess that's what these people want to say, too."

When interest in tourism gained momentum, Commissioner Hodgson of the Northwest Territories engaged Atlas Aviation for a flight to the Pole. And just to make sure the trip got plenty of publicity, he was to be accompanied by a Vancouver reporter. Such an opportunity seemed good for the tourist business. Weldy planned a preparatory flight to set out the essential radio beacon and extra fuel. On 5 April, 1971, a couple of days before the official trip, Weldy, along with co-pilot Jack Austin and Fran, set out for the Pole. They arrived safely, and placed the beacon and fuel – and during their brief stay brewed a pot of tea at the top of the world. They also planted for a short term, a flag from Hilson Avenue Public School in Ottawa, Weldy's earliest alma mater. That flag was returned for retention by the school, which in 1998, was demolished. Its demise was protested vigorously by many who remembered it with affection.

The trio returned to the base camp at Lake Hazen to meet the passengers. The reception was not quite what they expected. When Fran stepped down from the Twin Otter she was met by an irate Pat Carney, the reporter, who had assumed that she would go into the record books as the first woman to land at the North Pole. But Fran's record has been added to the Guinness Book of World Records, and is one of the achievements that can never be replaced.

The Hodgson-Carney party next fell victim to the capricious Arctic weather; they had to wait for the best of the often-unreliable forecasts for the polar site. When the report sounded reasonable they boarded the aircraft and headed for latitude 90 north. But when the aircraft reach the Pole, the ice was covered by fog, and Weldy could not land. The frustrated visitors returned to the south having only been over the Pole but unable to touch down. While Weldy's plan was to generate business by providing the complete polar round trip package it didn't quite work out. Instead, he was to provide support services for a variety of adventurers who wanted to trek across the ice. Some cases were a bit bizarre.

There was the Australian with his own helicopter whose variety of adventure was to fly to unusual places, jump out of his helicopter and videotape his accomplishment by holding the camera at arm's length and interviewing himself. And then there was the Japanese visitor who wanted to go to the Pole on a trail bike. A diminutive

German woman wanted to make the trip on snowshoes. Two Frenchmen wanted to test the Arctic elements in an ultralight aircraft.

In February 1969, adventurer Ray Munro, then 47, wanted to make his 500th parachute jump his last. The landing target he chose was the North Pole. The pilot he chose to drop him was Weldy Phipps whom he described as the best pilot in the north.

Munro was a former Spitfire pilot, flamboyant newspaper-man, and no stranger to radical ideas. He discouraged sport parachuting despite his own affection for it. His objective in the polar jump, in addition to capping his parachuting career, was to provide some information on the effects on the human body of such a cold descent.

When it came time for the jump, the Atlas aircraft was readied. Fuel was cached on the ice and clearly marked with fluorescent red balloons. But the poor visibility that stalled the jump also obscured the fuel cache. As the weather deteriorated, Weldy was forced to return from the drop zone with what fuel he had on board. Accordingly, he poured all available fuel into the tanks. Munro later told a newsman that he had emptied two and one half drums of regular gas into one drum of fuel oil then added 42 ounces of cognac. He declared that a drunk could breathe into the Twin Otter's engines and make it go. (The old reporter's talent for a good grab line had not waned.)

Munro made his jump from the plane at 10,000 feet. But he had to settle for a landing spot 120 miles from the Pole. He came down on an ice floe at 87° 30' north, 62° 25' west. On the way down he reported the temperature, with windchill, at minus 177°F. Having to remove his goggles on the way down, the icy wind froze his eyes shut. For Weldy Phipps the mission was not all that unusual in spite of a damaged nose ski for the first of two landings.

Fascination with the North Pole often got labelled as madness. Just such a term was applied to an expedition by Weldy Phipps when he was asked if he could support a 1967 over-ice polar trek. Ralph Plaisted, a well-heeled insurance executive from St.

Paul, Minnesota, broached the plan for an eight-man party. He made a preparatory flight to look over the land territory, and on the return from Eureka to Resolute asked Weldy if he thought they could make it. Weldy's quick answer was that the ice would likely turn them back. But he added that he was available for just such madness.

Weldy took on the job and supported the expedition to within 300 miles of the Pole where the group had to call it quits on May 4th because of the spring break-up – the ice Weldy had described. Pressure ridges 50 feet high as far as the eye could see forced the termination of the trip.

Undaunted, they regrouped, evaluated their miscalculations of 1967 and set out again successfully in 1968. The second trip started a month earlier to avoid the ice break-up of the previous year. Five members of the original expedition signed on again. The group included Jean-Luc Bombardier of the Ski-doo family. The snowmobiles he provided in the second year were much more powerful than in the first attempt.

Wiser in the second trek, the party made it to the North Pole and set the record for the first overland passage to the top of the world. They were guided by the waypoints Phipps established using SARAH beacons, smoke, rockets, mirrors and radio back to base to locate the trekkers. But the dreaded pressure ridges were still there.

Walt Petersen, mechanical engineer with the Plaisted group, told of his experiences at the expedition's 30th anniversary party held at St. Paul in 1998:

"Fly? I had never seen anyone handle an aircraft like Weldy did and I haven't seen one since. We were about 200 miles out on the Arctic Ocean ice. Weldy had set the Twin Otter down in his normal fashion, taxied up near the tent, shut down, off-loaded fuel, foodstuffs and snowmobile parts, went into the tent, had a cup of coffee, a cigarette. Then, back into the Otter, fired up the engines, backed out away from the tent and taxied to the east then turned west ... opened both throttles full, then cut her loose.

"As the Otter bounced westward, the props grabbing for air, Weldy saw an ice hummock about 18 feet high, dead ahead. He continued dead-on with full throttles. I retreated, thinking I don't want to get hit by flying debris from the plane as it crashes into the ice hummock. Then, to my surprise, Weldy brought the Otter up sharply just clearing the ice hummock. And no sooner had he cleared the hummock Weldy mushed the Otter on the west side of the hummock. This maneuver increased his airspeed only to now faster approach an ice pressure ridge. That, too, Weldy hopped the Otter over and he kept hopping these ridges until his airspeed was sufficient to maintain flight above all this mess."

Walt Petersen said he found Weldy to be a very capable man in whatever he did. "He knew how to work with anyone. He was honest, he was sympathetic, and he had respect for our heavenly Father. He was thankful for what ever was done for him. Weldy always respected others' for their talents or abilities."

One of the members of the first expedition, the late Charles Kuralt of CBS described Weldy Phipps as having an unusual relationship with an airplane. In a subsequent book entitled Journey to the Top of the World he described his skill landing on the sea ice by reversing the props of the Twin Otters just before touchdown, effecting a very short roll. On takeoff, Kuralt noted, he was meticulous about loading and often just cleared ridges by a few feet.

~~~~~

### The royal flap...

There has never been such a small and urgent shipment to the High Arctic as was made when Fran Phipps faced the dilemma of meeting the queen in 1969. White gloves had not been included in her Arctic kit. To beat the deadline forced upon her by Her Majesty's imminent arrival at Resolute, Fran sent out an SOS to Noreen Deacon in Ottawa. The gloves were sent northward by commercial jet and, just in time, graced a pair of kitchen-reddened hands for the royal presentation.

Having practiced her curtsy, she wondered what she might say to the visitor. When the time came, Fran was introduced to the queen whose only comment was, "Mm-m-m-mmmmm. When Weldy was presented, Her Majesty said, in a quick monosyllabic burst, "n-wots-yer-job-ere". At least that's how it sounded to Weldy who said something like,: "Huh? " Finally, he understood that the queen was asking what he did at Resolute, and told her he was a pilot. "M-m-m-mmmmm," she acknowledged.

~~~~~

## Politics

For a man who preferred to stay out of the limelight, Weldy Phipps was an unlikely political candidate. But in 1971 he stood for election as Northwest Territories Councillor for the High Arctic and won the seat. Apart from his preference for an isolation most politicians would decline, there could have been no better qualified candidate. By 1971 he had visited more places in his constituency than anybody else. He had developed a firm rapport and sound understanding of his 2500 constituents, most of whom were Inuit, 70 per cent unemployed, and trying to survive in a modernizing world by hunting and fishing.

Councillor Phipps promised to visit them four times a year. It was not a difficult pledge to keep since his company had been responsible for opening air access to most of the communities in the eastern Arctic. When Atlas first began to operate out of Resolute in the early sixties, some communities saw a visiting aircraft only once or twice a year. By the seventies the same communities had regular air service twice a week. In addition, the settlements could summon emergency airlift at any time.

The council seat gave Weldy an excellent platform from which to air major complaints on behalf of his constituents. He deplored the practice of allowing natives to sit idle while "rich college kids from the south" were given menial jobs in the north. He charged that those people arrived in the north for three or four months and then returned south with a fat pay packet. He gave as an example the presence of survey parties in which only the instrument man was required on a technical basis. The natives could pound stakes into the ground for them, and do all other support jobs but were shut out by the southerners. He wanted more salary money left where it could finance the natives in their hunting and fishing activities.

He urged the establishment of game reserves in the north in which no hunting would be allowed. Natives trained as game wardens could be employed to work in the areas, he said. Atlas Aviation had already proved that natives could be integrated into non-traditional work. The company employed two Inuit pilots and was training another as a mechanic at the time.

He was most critical of the federal Department of Transport which did not insist on the local hiring of non-skilled workers. Of some 60 jobs at Resolute, only five were filled by natives. White supervisors, Weldy claimed, treated the Inuit like dogs and then scorned them as unreliable. To illustrate his contention that the natives could hold down responsible jobs, he cited the case of an area supervisor who had gone south on a month's leave but whose job had been efficiently carried on by an Inuit assistant.

Weldy's term a councillor extended in its final year into his retirement. He had moved to Prince Edward Island but retained his seat until its expiry in 1974.

His affection for Arctic natives was well appreciated among the Inuit. When he left the Arctic he was presented with a narwhal tusk engraved with the names of several natives in the syllabics of their Inuktitut language. Weldy could identify each signatory by the inscriptions carved into the tusk.

~~~~~

## Chapter 6
## That southern yearning

After a few years of Arctic life, Fran Phipps had an uneasy feeling that she and Weldy would be living in Resolute forever. She was convinced that they would never be able to sell Atlas Aviation and retire. But just when her hopes for a less hectic life in the south were at their lowest ebb, Kenting Aviation came along in July 1972 with an offer that could not be refused. When the deal was settled the Phipps family could look to the future with a bankroll pushing the million-dollar mark. And just to put the stamp of approval on the deal, Kenting named the Resolute airport Phipps Field.

On the trip back to the south, Weldy was handed a present by the captain of the aircraft. Pacific Western Airlines issued an attractive certificate to anyone having crossed the Arctic Circle. The certificate confers upon recipients "The Arctic Seal".

The certificate to Weldy Phipps, however, was modified to read:

---

**This is to certify that our passenger**

*W. W. Phipps*

has been <u>SOUTH</u> of the Arctic Circle

---

**What now?**

Or better still, where now? Weldy had no shortage of friends. In fact, his ability to accumulate friends had borne him through the thick of war, the rough and tumble of bush flying and aerial surveying, and entrepreneurship. And it was an old friend from the Spartan days who set him on his post-Arctic course. Jim Wells, former legal counsel to Spartan Air Services, had a vacant home at Alberton on Prince Edward Island. He offered it to the family as their first refuge from the wintery blasts of Resolute Bay. They quickly accepted. The location suited Weldy's need to unwind and quickly wind up again with two of his long-suspended passions – machinery and boats. It would serve as a suitable base for Weldy's innovative ideas and the family's plans to explore the seven seas.

Sailing had been an early interest as was an affinity for things mechanical. The family scrapbook shows Weldy as a small boy holding a toy sailboat he had made. Later he had joined Ottawa's Britannia Yacht Club and learned elementary sailing in club races, and owned a boat on which he built a cabin for cruising on the Ottawa River. But around P.E.I. there was a lot of salt water waiting to be explored. And that called for a boat considerably larger than a dinghy. The choice was a 48-foot Hughes North Star yawl, which he named Whiskey Papa. It was delivered in Toronto in December, 1972, not a very good time of the year for a voyage home. The vessel was loaded onto a float and trucked from Toronto to Oxford, Maryland. Weldy and three sons – Dave, 21, Bob, 19, and Jim, 17 – flew to Maryland where the boat was commissioned. Then, armed with dinghy experience and Weldy's freshly minted Master's ticket, they set sail for West Palm Beach, Florida. Fran, with the three girls, Debbie, 14, Brenda, 10, and Terry, 4, drove to Florida.

Whiskey Papa drew seven feet of water, a definite handicap in the shifting underwater profile of the Intracoastal Waterway running down the eastern seaboard of the United States. And complicating the inland voyage was the frequent presence of bullying barge traffic. Anxious to get on with the southern voyage the green crew chose the open ocean for much of the passage. For subsequent trips on the high seas they engaged experienced blue water sailors to fill out the crew. Over the years, the boat underwent a series of recommissionings or upgradings. The original colour was a snot green. That was soon transformed into a royal blue with red trim – to match the livery of Atlas aircraft in the north. Weldy had chosen the dark blue on his aircraft to melt frost from the wings more quickly, if and when the sun appeared. There would be no need for frost protection where Whiskey Papa was heading over the next two decades.

But he wasn't finished tinkering with the original boat. The 40-horsepower engine, suitable for harbour maneuvering, proved inadequate for some of the sea conditions and currents encountered. One incident in 1983 proved the need for a heftier engine. Leaving a Caribbean dock for another ocean passage it became evident that an incoming freighter was headed for the same berth Whiskey Papa was vacating. Weldy ran the engine at full power but soon realized that it would not get them out of the way of the big ship fast enough. As their courses closed it was safe to assume that the ship's pilot would not be able to see Whiskey Papa. The situation quickly moved into an emergency demanding fast action by a well-disciplined crew.

Enter Red O'Neill, fellow prison camp mate and veteran landlubber. Red had arrived with great enthusiasm in October to sail for the winter. He was a 'character' who had amused himself in prison camp by tormenting Weldy with practical jokes. They remained great friends in the aftermath of their prison ordeal, and in spite of Red's ineptitude as a sailor, he was a welcome entertainer on board.

As the large ship bore down on them, Terry was ordered to ready the mainsail. Red was to man the winch that would raise the heavy sail and speed the departure. Terry had not yet snapped the halyard to the peak of the sail when Red hauled away on the winch. Up went the unattached halyard – no sail. Terry scrambled up the ratlines to try and retrieve the flying halyard. The exaggerated motion of the mast made her job dangerous. Weldy pushed the small engine to its limit as Terry snatched at the line each time it whipped past her. Finally she caught it, jumped to the deck and snapped the shackle to the sail's peak. The sail was hoisted. Whiskey Papa came to life and narrowly squeaked past the incoming ship. At his first opportunity Weldy removed the weak powerplant and replaced it with a 70-h.p. diesel.

Terry later described Red as a shipboard disaster. Her father was a bit more charitable. On an ill-fated voyage to Europe Weldy said that Red could not remember to keep a grip on something while the vessel pitched. He got thrown across the cabin, breaking some ribs and injuring a kidney. "I could have taken care of him with some painkillers," Weldy said, "but he wouldn't have anything to do with a needle." Red explained that the only thing he had ever used a needle on was his dog, and it died.

The trip to Europe got turned back in mid-ocean by a vicious storm that tore the sails from the mast. It was one of at least three storms Whiskey Papa was to endure and imposed a great respect for the power of the sea. During another storm the sails were again ripped and could not be properly taken down. The only way to reduce the hazard was to climb the mast and clear the snagged mainsail by hand. David Phipps was sent aloft in the wildly pitching swells. The boat was wallowing so badly that David was almost hitting the ocean's surface on alternating rolls of the vessel from side to side.

Not all lessons were so hazardously learned. Fran and the other neophytes had more than a few hilarious experiences learning the ropes – literally. Fenders were to be deployed before the hull bashed a jetty. Commands from the helm were to be responded to even when sound never travels from one end of a boat to the other. Lines had to be thrown ashore to bystanders even when they were too short to be of any practical use.

The captain was not excluded from hard-knocks training. Watching Haitian traders clean their hulls after beaching their boats at low tide, Weldy decided the same could be done for Whiskey Papa. And so as the tide ebbed the boat came to rest on the bottom and lay over on its side. The crew got to work scrubbing the accumulation of hitchhikers from the bottom. And then an off-shore wind began to blow. It took little time to realize

## Whiskey Whiskey Papa

that those poles sticking up through the sand were the masts of small boats caught in exactly the same conditions.

As the crew awaited the inevitable, the tide began its leisurely return. The boat began to show slight buoyancy. But not enough. To speed the process, the inflatable dinghy was launched, carrying an anchor to deeper water. The anchor was dropped in preparation for a kedging procedure. This required attaching the anchor line to the mast to right the boat and winch it back into safe waters. By now, however, the wind had raised bigger waves. Debbie was at the bow of the dinghy as it bobbed violently in the surf. When she turned to yell something to her Dad, he was not there. A wave had swept him overboard. Slowly, with her hull half scrubbed, Whiskey Papa made it back into deep waters. The chastened captain and crew had graduated to a new level in seamanship.

Whiskey Papa carried a variety of anchors for various bottoms. No anchor prepared the crew for one incident in the Intracoastal Waterway. The boat became stuck in the shifting bottom after being elbowed out of the way by barge traffic. An attempt to kedge the yacht off the bottom by winching to its anchor failed, and the U.S. Coast Guard had to be called to the rescue. A powerful cutter secured a line to Whiskey Papa, and attempted to haul the mired yacht out of the mud. There was no detectable movement at first. Then, with an explosive release, the embedded anchor let go, its braided dacron line recoiling and casting the anchor high into the air. It came down with a crash right on the deck, narrowly missing the cockpit crew. Weldy's disdain for the ICW was confirmed. Braided dacron and other ropes were learned along with a lot of other marine lore as the family plied the seas between Florida, Bermuda and Atlantic Canada.

A bold inscription in one of Whiskey Papa's log books says: WAKE WELDY IF A SHIP IS SIGHTED! The note was entered in the logbook during a voyage from St. Georges, Bermuda to Tortola in the British Virgin Islands. Bermuda was left astern at 1130 hours, December 4th, 1978. The weather was fair in a 10-

knot south breeze. Whiskey Papa was making four knots but soon picked up to six. Crewman Bill Grey is recorded as having mal de mer but doing OK. The big ship was sighted 20 degrees off the port bow at 1915 hours the next day. Weldy might not have been all that keen to be wakened because his turn on the helm came at 0300 hours the next morning. On December 11th at noon the yacht arrived at Road Harbour, British Virgin Islands. The trip is concluded in the log book: "Beautiful trip. Thanks to the Phipps and Whiskey Papa. God bless: Happy holidays in BVI. Affectionately, The Crew."

The return trip to Bermuda was started April 2nd. Six days later the log records "bloody cold!" weather and still 230 miles to go. They docked in Bermuda at noon April 10th.

Christmases aboard ship were spent in various warm ports of Florida, the Bahamas and the Carolinas. The relatively cramped quarters required ingenuity. The tree, for example, was fashioned from that part of the mast that was stepped into the hull in the salon of the yacht. The dinghy would be dispatched to harvest the branches of the Australian pines along the shore. These were attached to the mast in the cabin to resemble a Christmas tree. Decorations were fashioned from such suitable ornaments as ear rings and brooches. The stockings that were hung up Christmas eve were not exactly stockings. They were plastic bags acquired with groceries. They were shaped like stockings and stuffed in the night. Christmas dinners were restricted by the space problem. The turkey was more likely to be several cornish hens or a ham.

The galley in Whiskey Papa was another facility that did not meet the captain's approval. To make it more convenient, Weldy rebuilt that entire section of the yacht's interior. A master woodworker, he fashioned shipboard joinery of the finest quality. The new design made food handling at sea more manageable.

Winters in the warmer climes were interspersed with life on solid land back in Prince Edward Island. The family bought a plot of land with a woodlot at Alberton. This allowed Weldy's

## Whiskey Whiskey Papa

technical instincts, restricted only slightly by the years at sea, to blossom. He built a sawmill to produce materials for a log cabin. A large garden was cultivated and the produce canned. This sustained the family over one winter they decided to sit out in Alberton. Whiskey Papa was hauled out on the local railway ramp and secured for the winter. Weldy then built a huge workshop and filled it with machinery acquired from every heavy equipment auction they could find in the Maritimes. His shop was about the best equipped in Atlantic Canada. Weldy got used to the various new toys to search for alternative energy sources. No notable results were achieved.

When winter set in that year at Alberton, it seemed that the fates had objected to Whiskey Papa leaving her home at sea. Ice destroyed the rails of the Alberton Harbour and the yacht lay stymied in the large shed Weldy had built to accommodate her. But he was not to be thwarted. He ordered new rails from Nova Scotia and installed them at his own expense. This got Whiskey Papa back into action in local waters from a base at the Summerside Yacht Club.

In the summer of 1980, Whiskey Papa was entered in the Round the Island yacht race. Ten boats took part in the circuit which measured 350 nautical miles. Whiskey Papa came first, after 58 hours and two minutes, six hours ahead of the next fastest yacht.

The Phipps family soon learned that summers in P.E.I. were too short for their plans to commute back and forth to the Caribbean. Thoughts turned to the south. The Canadian property was sold to daughter Wendy in 1984, and Whiskey Papa emigrated to Bermuda. The St. Georges Yacht Club afforded an ideal base for meanderings around the warmer climes of the Atlantic and Caribbean. Fran and Weldy became active members of the club. Whiskey Papa rode at her mooring within sight of the new home.

Early in their residence it became evident to Weldy that the Yacht Club was in trouble. Its foundation, mainly steel pilings driven into the harbour bottom, had corroded to a point where it was feared they would give way and dump the club house into the ocean. Weldy scented a challenge, but by now he was wary of plunging physically into a solution. And so he advised the club members that he would show them what would fix the problem and supervise the necessary work. But he would not perform the labour. He conscripted the help of other Canadians for the work.

Weldy's remedy was to realign the weakened pilings and fill them with cement. Soon the building was declared stable and the activities of the club re-established. Such efforts gained Weldy great respect around Bermuda. One visitor marvelled at the typical pub crawl with him; "He never had to pay for a drink." The Bermuda residence lasted about three years before they cast off for the Turks and Caicos and Virgin Islands, then Florida.

With several years of island passages the younger Phipps crew had found their niche in life. They left the captain and first mate to content themselves with overnight trips back and forth to Grand Bahama from Florida. But if the short summers in the north posed a problem for sailors with wanderlust, the blustery summers in the south posed a more hazardous problem. The solution was to haul the boat out in Florida during the hurricane season and look for adventure ashore. Adventures such as it was, consisted of camping trips throughout the United States and Canada.

~~~~~~~

## Chapter 7
### Example by challenge

**W**eldy Phipps was obsessed with challenge. It was a characteristic that could have caused disaster if it had not been carefully channelled into productive endeavour. Nothing was impossible, he thought, if you believed you could do it.

The challenge philosophy was one he extended to his large family. He wanted to toughen Jimmie and Bobby to the rigors of the Arctic. The test he proposed would prove to the youngsters, pre-teenagers at the time, that they had the intestinal fortitude to meet any challenge. The opportunity arose in August, 1971 when he had a contract that would allow him to deposit the boys on Axel Heiberg Island's Gelman Glacier. The plan was to have them look after a supply camp for three weeks. It was to be a fuel cache for helicopters supporting a seismic project for oil exploration

Their mother was worried that a polar bear might get them. "No problem with bears. They'll be OK," Weldy assured her. But he armed them with a .303 rifle and 100 rounds of ammunition.

"They might wander off and fall into a crevasse," she said.

"I told them not to wander too much," Weldy said. The assignment went ahead.

Their first job was to spread bags of sawdust as a bed for a 50-foot long rubber bladder which would hold fuel for a helicopter operation. The Atlas DC3 would fly fuel to the camp, landing on an uphill surface. The boys would pump fuel from barrels into the bladder. They were also required to keep a landing strip open for the DC3.

For creature comfort the boys had a supply tent and a double-walled teepee for sleeping. The living tent had a small stove. Weldy had instructed that the stove was to be refilled outside the tent. But when weather closed in and dumped three feet of new snow, they decided that if they were careful the stove could be filled indoors. When it was lit, however, it set fire to the tent. The flames were extinguished with snow, flour and oatmeal. But that lesson with fire did not last long. A little later they forgot to put the propane stove out at night. It toppled over setting fire to Bobbie's cot and sleeping bag. The pair struggled out through the tent's tunnel entrance with the burning material.

During their isolation, the boys used the camp stove for warmth. It was comfortable enough. But when Bobby noticed Jimmy becoming incoherent he realized that the oxygen in the tent was being used up by the stove. He hustled his brother out of the tent to let the Arctic winds revive him.

At first, food was no problem. Condensed rations were common with Atlas crews since various expeditions would leave their surplus supplies with them when they left the north. Such rations, however, were supplemented from Resolute with such amenities as freshly-baked pies. But a late demand for food returned only a case of peaches, a case of chop suey and a bundle of tea. Then the weather closed in, cutting off visits of the DC3 and new supplies. The lonely (and only) inhabitants were stuck there for several days. In the long meantime, they were on the radio to their mother at home base. "Tell Dad to hurry up," they pleaded. "All we've got left to eat is peaches -- fried peaches, frozen peaches, chop suey, and peach tea."

~~~~~

## Wide-eyed but flying blind...

David was 13 years old when he spent his first summer in the north. He had been attending Albert College in Belleville, Ontario, and had yet to face the challenges his father had waiting for him.

"I remember the long flight from Montreal to Frobisher and on to Resolute aboard a DC-4," he recorded many years later. "The trip was about 13 hours flying time and passengers sat in seats behind the cargo section. The seats were not particularly comfortable, especially for sleeping. Rather than sleep in the seat, I took a blanket and pillow and found a comfortable place to lie down atop the cargo. When we arrived at Resolute the base was blanketed in fog and we ended up diverting to Hall Beach until things improved. What a disappointment. The following day we got a break in the weather and made it in.

"Upon my arrival I was met by father and we set off for the trailer complex to get me settled in. My first chore was to clean up the kitchen. There wasn't a clean dish in the place. I was not amused. It took me about two hours to clean the place up and put away what seemed like fifteen complete place settings of dishes. I did such a good job that I was the designated galley slave for the rest of the summer.

"The northern experience was something I will never forget. Travelling to areas where you are the first person to set foot on or even to see was quite a feeling. The vast barren landscape seemed endless and its beauty and peacefulness overwhelming.

"Much of that summer I spent flying with my father in his brand new single Otter. I don't know where he got his stamina, but, we would fly for hours and hours and just when I thought we would get a break, we were off again.

"On one particular trip to Frobisher Bay my father was getting weary and asked me to take over for a couple of hours while he went into the cabin to get some rest. He had a couple of thick cushions he could stack up on the seat to help me see out the windshield. Well, a couple of hours passed and I looked back into the cabin and saw my father sound asleep and didn't have the heart to wake him up.

"Everything was going okay. I had my course to fly and altitude to maintain and all the gauges appeared to be reading normal. I had to make a couple adjustments to the trim tabs and basically the plane would fly itself. As time passed the weather deteriorated and visibility was reduced to almost zero. I kept my course and altitude for a couple more hours after which my father woke up and climbed back into the cockpit. He was surprised he had slept so long and even more surprised that I had been flying on instruments for the past couple of hours. He checked our position and found that we were pretty much on track. He asked that I be sure to wake him up whenever I get into weather like this again.

"I was with my father as he executed landings and takeoffs in many situations that made me hold my breath and hang onto my seat with white knuckles. I would look over at him and he would be as cool as a cucumber, his concentration focused intensely on the task at hand. On one occasion we had landed at Grise Fiord in high and gusty wind conditions.

"We unloaded the aircraft, and I noticed that the wind gusts were almost picking the single Otter off the ground. I pointed this out to my father and he didn't seem too concerned. We loaded the plane and prepared to depart. We had to take off in the direction of a nearby mountain which didn't appear to be too far off. I was getting concerned but had great faith in my father's ability. I swear our take off roll was no more than 10 feet before we were airborne. We seemed to climb without any forward movement and then we veered off away from the mountain and we were on our way back to Resolute.

"As I flew around the Arctic with my father I realized how he could have fallen in love with it. Although the conditions were very harsh at times, he enjoyed the challenge and solving problems that others would call impossible. The Arctic landscape was also something to behold. The beauty of the virgin landscape and the sense of being totally alone and out of touch with the bustle of modern society was almost a religious experience.

"We often flew to areas where early explorers had been forced to set up winter camps, their graves and buildings preserved by the cold temperatures and untouched by archeologists. We would discover cairns that held sealed metal tubes containing maps and messages for those who followed, whether it be a search party or a visitor. We would leave our own notes giving details of our visit. It was experiences like these that drew me closer to my father and made me realize why he became a pioneer in such a barren land.

"My first summer in the north with my father was the one I remember the best. The time we spent together was a non-stop learning experience of work ethic and values. Much of what he taught me that summer remained with me and shaped my future. Although we had our differences from time to time he always had my utmost respect. I loved him dearly and miss him every day. He accomplished more in one lifetime that most men could ever dream of accomplishing in three. I am very proud of him and proud to be his son."

~~~~~

*[David died in September, 1997, at age 46 of lung disease. He had worked at one time installing security systems. This work often required placing wires through attics and other confined spaces. On one occasion the workers were visited by fire inspectors dressed in full protective gear including masks. Alarmed at the presence of the installers without protection against fine particles of asbestos they said that a fireman would not even look into such spaces without a mask.]*

~~~~~

### Janet: a chip off the old block...

Janet, first of the Phipps children, lives in Surrey, B.C., and works as a Registered Nurse.

"My first vivid recollection of my father goes back to Ottawa when I was a small child. Dad worked at Spartan Air Services and on weekends we would pack up picnic lunches and head to the airport to watch him fly or wait for his arrival from some faraway place. I remember my mother saying: "Here he comes now." Gazing skyward, I could first hear and then see the small aircraft — a shiny speck in the sunlight, then looming larger, circling around and above us and tipping its wing in a loving greeting. I recall the butterflies in my stomach, tears streaming down my cheeks and my hear bursting with pride as he landed safely and taxied towards us.

"I loved my father's smile and the smell of him as he hugged us. He always had huge chocolate bars which we ate with gusto even though they were mottled with age, and tasted of grease and oil. 'Part of my survival kit,' he would explain.

"When Dad worked at Bradley Air Services some years later, he engaged my sister and me to work in the canteen making hamburgers and hot dogs. He paid us $5.00 a day and good-naturedly grumbled that we were making more than he was.

"My father worked long hours and spent much time away from home while I was growing up. He would explain that this was necessary in order to build his business and provide a good life for his family.

"When I was seven years old, Dad bought me my first two-wheel bicycle. As he adjusted the seat and reviewed the rules of the road, he recounted on of his When-I-was-your-age stories. 'You are very lucky,' he said, 'When I was a boy we were so poor that I had to scrape rubber from the roads of Ottawa in order to collect enough to make bike tires for my homemade frame.' Dad had many When-I-was-a-boy stories, embellished no doubt, to teach us the value of a dollar and the merits of hard honest work.

"As the 1960s came into full swing, off I went to join the fun — travelling around the world and behaving most hedonistically and irresponsibly. This bewildered and displeased my father but I know that deep inside he admired my adventurous spirit. After all, he would coyly admit, years later, that I inherited that trait (as well as all of my other positive ones) from his side of the family.

"In 1970 he sent me a photo from Resolute Bay of him and my mother dressed up as hippies, en route to a party at the Arctic Circle Club. He made an hilarious picture of a cross between Sonny Bono and Liberace.

"After my parents retired and set off sailing on the Whiskey Papa, I didn't see them for several years. In 1985, I was working as a registered nurse in New Brunswick and enduring a painful divorce. I called my father in Florida and flew down to spend six weeks on the boat with him and my brother Jim.

"As I nursed my wounds and made plans to restructure my life, my father quietly offered support and advice. After a particularly heart-wrenching talk, Dad got up to mix us each a drink. As He walked away I heard him mumble to himself, "As far as I'm concerned, he was never good enough for you anyway." In his own way, he was reassuring me that people come and go in a lifetime but a parent's love endures forever.

"After Mom and Dad moved from Bermuda to Texas, Dad and I began to communicate more regularly. His telephone conversations and letters were poignant and witty even as his health was failing. Dad was the most intelligent, inventive, self-effacing and honest man I have ever known. He was diligent and generous to a fault, always the champion of the underdog. Most of all, my father never settled for second best — and that's why he married my mother. I miss him greatly and I am very proud to be his daughter."

~~~~~

## Terry, confessed bilge rat...

Terry was three when the family left the Arctic. "My memories of my father begin in the second phase of his life – his years on the sea, not in the sky. When we moved onto the boat, I was six years old. Less than two months into my first year of grade one. I had no idea what adventure lay ahead of me, or how it would impact my life. I would be on correspondence courses for most of my elementary school years as we explored the Gulf Stream, navigated the Intracoastal Waterways, surveyed remote Caribbean Islands, and tempted the Bermuda Triangle with stubborn regularity."

Secondary school continued when the family returned periodically to Prince Edward Island. Educators agreed to accept the students at their supposed level so long as they could keep up with the curriculum. None of the seaborne correspondents ever had to be class demoted.

"I never remember feeling fear when we were 500 miles out to sea with the wind whipping fiercely through the rigging and the ocean pounding against the hull of the boat. I never doubted Dad's ability to get us safely to our destination. But I never found my father's skills remarkable. Not until I was older and realized that not every man possessed his knack for navigation, mechanics, carpentry – you name it. The wheels in his head were always turning.

"In the evening he would sit quietly at the table with a pencil and a piece of paper, drawing diagrams of how he was going to repair, improve, or invent his next project. He could fix everything out of anything. But this is part of Weldy's legend and, although remarkable, it is not all of what defines Weldy as my father.

"Some of the everyday occurrences of our life on the boat form many of the fond memories that I have of Dad. I held the title of 'bilge rat' because, being the smallest, I was sent into the bilge to retrieve tools that he had dropped. I would free dive off the side of the boat to fetch tools and glasses that had fallen overboard. I felt very triumphant when I surfaced from the depths of the sea, or the bilge, with a wrench on the end of my magnet.

"Dad did not shelter me or exclude me from things because I was a girl or I was too young. He would give me small challenges to take on, pushing the expectation of a child of my age. For instance, because I was light and small I would be sent up the mast in the bosun's chair with instruction on how to repair the masthead light or where to insert a bolt. I will always rise to meet a challenge because he taught me that challenges enrich our lives and help shape our character. And, more through deed than word, Dad taught me that through original thinking and determination, I can do anything that I put my mind to.

"One very fond memory I have of Dad is when I was 10 years old. We were spending some time in the Virgin Islands. By this time I had already learned the basics of how to use sextants to navigate by the sun and stars. My knowledge of the stars was limited so my father endeavored to teach me the constellations and how to orient myself in the night sky. This absolutely fascinated me.

"I would wake up in the wee hours of the morning and creep out to the dew-covered cockpit of the boat. The air would be silent and still, except for the occasional slapping of the halyard against the side of the mast. Dad would point out the most common constellations, explain how to get from one to another and, most importantly, how to determine which way was north so that I could determine the direction that I was headed. It wasn't long before I could quickly point out Betelgeuse, the North Star, the Big and Little Dippers and the Three Sisters. There were many others that we learned and I felt very proud when I was able to look at a map of the constellations and then point out a new one to him.

"Dad passed on a determined and adventurous spirit to me, one which will stand by me and ensure my success in life. I am proud to possess it and I will endeavor to pass it on to my children."

Terry has succeeded, making good use of the commerce degree she earned from from the ubiquitous University of Maryland – ubiquitous because of its arrangement with the armed forces of the United States. At worldwide locations of U.S. forces the U of M maintains access to its courses. Terry was an employee of the U.S. Navy while the family was resident in Bermuda. She is the business management partner in her sister Brenda's veterinarian clinic in Calgary. In 1992 she married Bermudian Brian Bean, and is the mother of two boys.

~~~~~

### Debbie: Recalling the dynamic duo...

"It is with great pride and pleasure that we, the offspring of the most dynamic duo that I will ever know, have been asked to participate in the writing of a book about their lives. I feel it necessary to include my mother in any account I may write about my life experience and memories in regards to my father because they were an inseparable entity throughout my life.

"From my earliest memories where I can always hear Mom saying 'there is nothing your father can't do' until Dad's final moments in a hospital bed in Ottawa, the strength of their bond was one that was easily palpable as only a bond created by a life of bold adventure and compassion for change can conceive.

"I was born the seventh child to parents who at that time were just beginning a quest that would take us on an incredible journey that began in Ottawa and going full circle to end most appropriately for my Dad in the city of his birth.

"My earliest memories begin in a crude little hangar outside of Ottawa where Dad kept a little plane (which I believe was a Piper Cub) and a pony named Domino that he had given us for Christmas. I was too young to appreciate the plane and because Domino had a habit of biting everyone who extended a hand to pat him, I shied away from him. I did however find all the nuts and bolts and the various grease-laden airplane parts scattered over the work bench rather intriguing. It was in this hangar where I first learned about tools, never knowing that I would one day be expected to participate in dismantling a diesel engine in the bilge of a boat or a truck engine in a shop in Prince Edward Island years later. If he 'could do anything' then so could we, and gender had no bearing on the task at hand.

"Little did I know at this time that Dad was preparing to head to the Arctic in that little plane, so tiny that all it could carry was emergency supplies and a drum of fuel with Dad squished into the small cockpit.

"Over the next few years we rarely saw Dad. He flew all over the Arctic while Mom managed the ever-growing family and all the 'office work' from our home on Thorson Avenue in Ottawa. Dad's visits to the 'south', (which we all came to refer to Ottawa as) were rare in my recollection, perhaps because there were so many of us in the house or I was just too busy playing with my Barbies to notice. But I do remember the airplane parts in the basement in various crates and boxes.

## Whiskey Whiskey Papa

"When Dad was there he was always working on something, which I accepted unquestionably as what my Dad did. I had no idea of what anything was or what purpose it served. All I knew was he could fix anything and he always had grease under his nails. I often would look at my friends father's hands which were for the most part clean and manicured and dismiss them as someone much less important than my Dad because in my mind they were afraid to get their hands dirty therefore lacked knowledge and skills necessary to do 'real work'. I was yet to find out what that real work was.

"Early in my childhood, at the age of six, I was moved out of my familiar suburban environment to a new home hundreds of miles north in a world without days all winter and without nights all summer. Resolute Bay was a new world for all of us and the only thing familiar to me were my best friends who were my seven siblings and my parents. Little did I know that this would be a life that bonded us all together to form the tight knit group that still exists today.

My first night in Resolute Bay was like a camping trip. After a long flight north from Montreal on a DC-4 we disembarked on a tarmac that was in the middle of a treeless, cold environment. There was no terminal with luggage carousels and friendly faces welcoming you to this new destination, just a cold tarmac and Eskimos dressed in unusual outfits.

"My first impression as you can imagine wasn't one of sheer delight. I thought it was ugly and empty. Little did I know that years later I would board a plane and cry my heart out to be leaving my 'home' for the very last time. Once arriving in our new place of residence we collected our luggage which consisted of many cardboard boxes, a few suit cases and many bags of clothing and headed to our residence.

"The trailer was rugged to say the least with this one room which was our sleeping quarters. It consisted of a rather large room with big wooden shelves made of 2x4 lumber and plywood. Here we unrolled our sleeping bags and spent our first few nights sleeping on our assigned shelves in morgue-like fashion. It was certainly cold enough to be a morgue! But Mom and Dad made the whole experience an adventure which we all found exhilarating. This was the beginning of a childhood of adventure that has shaped us all into the adults we are today.

"Our living arrangements did improve over the years that we spent in Resolute. In the beginning we were housed in a set of three army trailers that were attached together. It wasn't anything fancy, actually it was what most people would consider rather crude. However, it was home to us and we maintained our family spirit despite the lack of modern conveniences.

"The most memorable daily tasks in those early days were centered around the most basic needs. The kitchen consisted of a propane stove, crude counters made of plywood sheets on top of old wooden crates and a huge table. There was no running water therefore no sink either. Mom cooked for the family as well as the entire crew in this makeshift kitchen. All the water was hauled from the base in steel milk cans.

"As kids anywhere we had our 'chores' to do. My brother Jim and I were responsible for transporting the 'honey bucket' to the dump. Although an outhouse is a more practical solution in a southern area, the thought of exposing your bare bottom in 60 degree below temperatures was out of the question. The honey bucket was our toilet that consisted of a plywood box housing a bucket lined with a green garbage bag. When it was full, Mom would set the bag outside and Jim and I would load it onto a sled behind the Skidoo and take it to the dump. It became the norm for us, and we even enjoyed our trips to the dump.

"Our bathtub was simply a huge old-fashioned wash tub that we could barely fit into. You had to sit in it with your feet and legs hanging over the edge while Mom heated water on the propane stove and dumped it in the tub. After a few years of hillbilly life we moved to a modern trailer on the base which was connected to the base water supply and sewage system. It was like heaven. I remember how excited we were to finally sit in a bathtub like people 'down south' did.

"School in Resolute (Qammartalik) was much different than my brief encounter with the education system 'down south'. We had two rooms, one which housed grades 1-3 and the other grades 4-6. After grade six it was either correspondence courses or boarding school where my older brothers and sisters were. Myself, my brothers Jim and Bob and my sister Brenda were the only white students in the school except for the teacher and village administrator's kids. The latter of which seemed to change from year to year.

"We were up at 5 a.m. and into a Bombardier, a tracked vehicle with skis that went to the village every morning at 6 a.m., to pick up the Eskimo workers which were employed on the base. We usually fell asleep on the way to school, it was easily achieved, given the absence of daylight and the drone of the motor. School was school, but the recesses and lunch breaks as well as the two hour wait in the morning and the two-hour wait after school were quite fun.

"We often went to our Eskimo friends' houses at those times where we were exposed to real Eskimo culture. The frozen seal laying on the cardboard sheet was dragged in from the porch and the whole family squatted in front of the carcass and cut off chunks of frozen seal meat. We looked on, and munched on peanut butter sandwiches.

"After a few years in the north the Eskimo meal times no longer seemed crude but rather quite acceptable and even practical. The sight of a mother chewing a bolus of raw meat and then spitting it into her hand and feeding it to the young baby on her back was just the way it was. Survival was the game and the Eskimo people were expert players.

"Dad was really busy during our years in the north often gone for days at a time. Sometimes when Mom wouldn't let us do something we wanted to, like go to the movie, we would just bypass her and get on the radio to Dad. 'Whiskey Whiskey Papa, this is YYZ, come in Dad'... crackle, crackle 'YYZ, Whiskey Whiskey Papa, go ahead'... more crackling... 'Dad can I go to the movie tonight'... crackle, crackle...'Uh YYZ, ask your mother dear' was usually the response! Darn!! I had no idea that not only Dad could hear us but also everyone else who was in radio range across the barren tundra. I am sure we provided moments of entertainment for the people who had chosen the northern experience.

"As a kid I had many opportunities to tag along with Dad on his many flights into settlements around the Arctic. Our aircraft were small and didn't have many of the fancy amenities that the larger aircraft had like bathrooms and galleys. But with our boxed lunches and 'pee pots' we managed well.

"I can remember a couple of flights with Dad where I was co-pilot. I couldn't see over the dash unless I piled log books on the seat. On a couple of occasions when Dad needed to visit the pot in the back of the plane he would leave me at the controls all alone and give me a point of land to steer towards while he was in the back of the plane 'conversing with nature' as he put it. When he returned I was proudly sitting forward, straining to see over the dash and knowing that I had indeed kept on course toward the landmark and would wait for his praise. 'That's pretty good Deb, but you just increased the altitude by about 1500 feet!!' So who mentioned anything about altitude???

"When I was about 11 Dad increased his fleet by buying a couple of DC-3s. They were luxury craft compared to what we were accustomed to. They even had a bathroom and a small galley. In order to improve customer service Dad decided that he would like to provide meals onboard and thus my short-lived career as a stewardess began.

"My training was pretty basic. Dad gave us a manual to read which contained a lot of boring statistics, as I remember. He pointed out a gauge in the cockpit which was an indicator for the level of hydraulic fluid necessary for operating the landing gear. I had no idea why he thought this was important for me to know about it, but when he said, 'If you ever see that below this level you need to add fluid, if there is no hydraulic fluid to add then you put anything in even if it means peeing into it'.

"For the first few trips I would cut a sandwich, check the indicator, put the sandwich on a plate and pour a coffee then check the fluid level again. I became so fixated on the column of fluid and was terrified that if I didn't check it often, we would end up doing a belly landing and it would all be my fault. Dad had a way of making his point clear.

"Family vacations were few and far between as Mom and Dad worked endlessly establishing themselves in pioneer-like fashion while raising a family at the same time. We did make a few treks to the south to visit relatives in Ottawa and restock on items that were unavailable through the Sears Catalogue.

"The most memorable trip was made in a Beechcraft. A rather nice plane as I remember that had much nicer upholstery than the typical cabins of Dad's fleet. But there was one small problem, the heater didn't work. We were all buckled in our seats in sleeping bags, mitts and toques. The air temperature on the ground was tolerable, but once in the air, the temperature seemed to plummet to the point where we couldn't feel our feet. I remember we were all singing and laughing and passing food around.

"We overnighted somewhere at an old abandoned base. It was a U.S. Weather Bureau base, as I remember, because for some reason I felt in necessary to steal a stainless steel mixing spoon, to this day that spoon is still in my kitchen with the letters USWB engraved on the handle.

"We reached Ottawa a couple of days later, then went on to Florida for a real vacation. Our northern unexposed skin burned to a crisp and my sister and I suffered for days while Mom bathed us in Noxema.

"When I was 13 my Dad sold Atlas and entered what he called retirement. Although he no longer owned Atlas, throughout his years of 'retirement' he continued to work a lot. We moved first to Prince Edward Island - a temporary move until our boat Whiskey Papa was built. We moved into an old log cabin on a river near Alberton. Although a new place, the experience was full of similarities to many places we have called home – the lack of running water, flush toilets and electricity. But in the usual Phipps adaptation mode we made out just fine. We stayed there for the summer before heading to Maryland to pick up Whiskey Papa and begin yet another new adventure.

"The boat was much smaller than I had anticipated but it was so pretty. None of us had a clue what anything was or how anything worked but we all did share a trust in our parents' skill and knowledge and knew that we would be OK. Over the next few years we learned everything that we had to know about the boat. There were no formal sailing lessons for us, we just learned by doing. Out of necessity, our roles and responsibilities were defined. Although the contrast to northern living was profound, there were so many parallels.

"We spent a lot of our time in remote places, we had to 'make do' with what we had. Mom is probably still the most resourceful person I have ever had the privilege of knowing. She could find a solution to almost any challenge and always maintained a sense of humour ... well almost!!! There were a few times when I thought she was going to launch the dinghy and leave us.

"Once when we were in very heavy seas and she was trying to make lunch. There was a pot of beans in the refrigerator that she wanted to heat up. She would wait for the boat to lean to starboard and try to open and close the refrigerator as fast as possible before the boat tipped the other way. She wasn't quite fast enough and when the boat leaned to port Mom still had the refrigerator door open and the pot of beans landed on her new deck shoes. I can't really repeat Mom's comments at that time, and it's not because I can't remember what they were!

"We spent a lot of time exploring the islands of the south during the winter months then returning to Prince Edward Island for the summer months. Our education was through the Alberta Correspondence School. It was really difficult to discipline ourselves to sit and actually get the lessons completed. There were just so many other things to see and do. We all became certified scuba divers in South Caicos and spent hours diving on the reefs. We met some really interesting people from all over the world.

"The education that we all gained from our life experience is tenfold what any classroom could have given us and the lack of formal education certainly didn't impair any of us from becoming successful adults.

"It was very apparent to all of us kids that our lifestyle was radically different from that of our relatives and friends. While there were times I yearned for a more 'normal' life like everyone else had, there was such an infectious sense of pride within our family. We are all survivalists in our own right and our parents taught us to look beyond a simple solution, to adapt, to resource and most important of all, to never give up. After all, if Dad could do anything, then so could the kids. The words 'I can't' just didn't exist.

"At the present time I am living in Charlottetown with my husband Darcy and teenage kids Arlo and Tara. I am a critical care nurse and dream of practising in the north 'just for the experience'."

~~~~~

## Coffee, coffee or coffee...?!

The initiation of two of the world's youngest stewardesses is one the candidates won't soon forget. Wendy Phipps was 16. Her sister, Debbie was 12 when they were dispatched by their mother to perform flight attendant duties on an Atlas Aviation DC3 bound for the Arctic drilling camps with a manifest of oilmen. Little did they know they would be providing the in-flight entertainment as well.

Their duties, as explained by their mother, was to serve coffee and sandwiches during the flight. The girls took their instructions very seriously and looked forward to the challenge. But when it came coffee time, the complaints from the passengers started to come in. One after the other told them:

"My coffee cup's leaking!" The girls ran with the dripping cups to transfer the coffee into new cups. But the new cups leaked just as badly as the ones they replaced. The complaints multiplied until the girls were run ragged trying to keep the coffee contained.

When all the coffee had leaked and the two stewardesses were resigned to washout status has hostesses, they were let in on the whole story. The co-pilot, Hans Hollerer, had put pin holes each cup just to provide in-flight entertainment for the passengers who were also in on the gag.

~~~~~

### Brenda: inspired by lemmings...

"To even begin writing about my father seems such a formidable task. Where does one begin to describe this man? I suppose to me, he has been one of life's great mysteries. I don't know anyone who has exuded such greatness and yet such vulnerability as a human being. Though he was not one to speak of his internal workings, there was never a doubt in anyone's mind as to his depth as a person. He was an ocean of thoughts, ideas and perceptions intertwined with as many personal struggles as any sensitive human being.

"As an adult I am thankful to have reached a point in my life where I can appreciate his qualities as a person, a genius, my father, to get a glimpse of those lives he has touched, including my own. As his child I am extremely grateful for the part of him that I am.

"There are many memories now from my early childhood days in Resolute Bay. We seemed to spend a lot of time exploring. Everything was an adventure and my brother Jim was usually the leader – full of ideas and imagination. One of our favorite locations was the dump where we'd spend hours taking colourful lights off discarded radio equipment.

"The best trips were when we found stacks of boxes filled with expired U.S. Weather Bureau balloons. Off we'd go to the hangar to inflate one with Dad's air compressor. The balloons could fill a small office or two once inflated, and on a few occasions they did. Mostly, we'd drag the balloons back outside to play. If you ran as fast as you could into one of those things, you could almost take flight on the rebound. Yes, these were Weldy's children all right!

"We had a variety of activities to occupy our time, depending on the season and school schedules. During the winter, we'd ride around on Skidoos, or have duels with the world's biggest icicles. Large cardboard boxes were converted easily into homemade Crazy Carpets. The plastic version always split from the cold, anyway. If I were feeling really brave, I'd join the older kids jumping into a snowbank from the roof of some carefully selected DOT building.

"In summertime, we really had fun with the freedom of warmer weather. There was an old plane wreck in a nearby bog, just off the base. With morbid curiosity, we'd imagine finding skeletons in uniforms inside, or someone's leg a few feet away. This particular wreckage, we later referred to affectionately as the 'Poo Plane'. We thought it just smelled kind of funny until Mom found out exactly where we had been, checked out our slimy footwear, and announced that it was located right on the sewage line.

"I remember lemmings everywhere we went in warmer seasons. It seemed such a harsh environment for the poor little things. Thus began my veterinary career – full of miraculous intentions. The lemmings were easy enough to catch, using my jacket as a net, then bunching it up to form a sack. At last I could rescue one and give it the warmth and care it so rightfully deserved. The only problem was that iceberg lettuce with a lemming in a large, sealed mayonnaise jar, placed beside the heat vent, apparently wasn't the answer. Fortunately, I was much older when I realized I'd suffocated the poor thing. At six years of age it would have been quite devastating.

"One of my favourite memories is of flying with Dad in his Twin Otter. Occasionally, he would let me tag along on a day trip from Resolute. The best part was sitting in the co-pilot's seat. It was so exciting with what seemed like hundreds of gauges to watch and switches everywhere, my feet dangling over the edge of the seat and my headphones loosely draped around my head – 'Roger, Dad' 'Over and out' – Boy, was I important!

"Eventually, he'd switch on the autopilot, unbeknown to me, and ask me to take over for a few minutes. I'd hang on to the two big handles in front of me, trying to maintain a steady course to somewhere I couldn't see from my perch a foot below the cockpit window. Now and again he'd say, nonchalantly, something like: 'Don't pull on that handle too hard, dear. You might crash the plane'. Then he'd calmly close his eyes again while I'd sit frozen with sweaty palms and little white knuckles. Boy, what a relief when he took over once again!

"Living on the ocean with my parents is certainly a big part of my past. Alas, another adventure. Every day was something new. Any time that we settled for even a few days Dad was busily repairing anything and everything. Since a boat required constant maintenance, it is perfect for someone with such an active and inventive mind. It didn't take him long to convert the navigator's bunk to a six-foot cupboard for tools and spare parts.

"He quickly developed a reputation everywhere we went as the guy who could fix anything. Nowhere did we see such a comical example of this than at Atwood Harbour in the Bahamas. It was a small, natural and remote bay on a small out-island with a very small town about 10 miles away. We were attracted by its good shelter and a supply of fresh water not far from the beach.

"Dad was constantly puttering on the boat with navigational equipment strewn in pieces across the table in the main cabin. He'd spend hours muttering to himself and peering over his latest project through his thick, black-rimmed glasses. His eyewear was a constant source of entertainment for us kids. He never seemed to notice the permanent pits in the lenses or the oily fingerprints from sliding across the cockpit or his hours working around the engine. Nor did he mind the bright yellow floats (purchased in a marine store) that we fastened to each of the arms. The glasses had explored the depths of the bilge as well as the ocean floor more than once.

"If he was awake, Dad was repairing, redesigning or inventing something new. One day, while at anchor in Atwood Harbour, Bahamas, we spotted a man in a small dory from the town a few miles away. He ventured over to say hello and eventually asked Dad to help repair his transistor radio – not much of a mystery, it simply required new batteries. The following week or so a few other projects emerged. A day or two after the radio repair, came the second job. This time it was a small outboard motor. Within hours, Dad had it in pieces on the deck and repaired in record time. No sooner was this job completed than the man arrived with a larger outboard motor. Of course, Dad could fix it. He was some kind of genius.

"We cannot begin to tell you how tickled we were to spot our new friend emerging from the distance a few days later with an old winger washing machine precariously balanced across his small boat. We set sail to explore new horizons.

~~~~~~

## The sixth sense

Jim Phipps thinks he knows why his father survived the odds that often tended to stack against him. "It was his sixth sense," said Jim, pondering several incidents in their travels together. He recalled one medical evacuation to Frobisher Bay (Iqaluit) in which the ceiling and visibility at the Frobisher end was reported to be zero-zero. Weldy continued, confident that an opening in the low overcast would occur. If it did not, there was no alternative but disaster. When he reached the terminal point, committed to land, a small clear patch opened for a few precious seconds. Weldy dropped the aircraft through the hole and landed safely.

This 'sense' was not restricted to the air. On another occasion Jim and Weldy were to meet at Lucaya, Bahamas, for a voyage on Whiskey Papa to Bermuda. Jim had not seen the yacht's new dark blue colour scheme. He almost boarded the wrong boat, but turned and decided to check elsewhere first for Weldy. As he climbed the stairs out of the lobby loud music was blaring. He reached a point from which he could see into the bar. When he looked down, Weldy suddenly turned and looked right at him. Jim credits Weldy's mysterious sense. And he feels that the same sense turned him away from the blue boat. It's cocaine smuggling crew were in jail. He might have been under the same suspicion if he had boarded it.

But that was not the last intuitive incident of the day. The pair set out on an evening departure from Lucaya. After nightfall they heard a radio call from a sailing vessel unsure of its position with respect to West End. Weldy made radio contact and sent up a series of flares for the other captain to see. No response. And then a white light, indicating the stern of a vessel, came into view. Whiskey Papa gained on the other boat and Jim could hear Weldy's voice coming back from that boat's radio.

Weldy went below and turned his radar to its widest reach. The return echoes indicated a circle of six other craft with the nearby boat heading away from them into the centre of the group. The position of the confused vessel was on the far side of the six. Weldy smelled a rat. He went into the cabin and came back on deck with a rifle and a handgun, sure that Whiskey Papa was being drawn into a pirate's trap. Two earlier incidents confirmed his suspicion that the radio call was a ruse. He resumed his course for Bermuda.

Earlier in the day, when Jim and Weldy had boarded Whiskey Papa they were detained by two men bearing submachine guns. A third, obviously American, toted a shotgun. They demanded to search Whiskey Papa supposedly for smuggled Haitians. Finding none, they mellowed and left. Later in the departure, a Coast Guard vessel had overtaken them and questioned them about their point of origin and destination. Whiskey Papa was cleared and proceeded.

Pirates anxious to get their hands on speedy sailing craft were not the only hazard on the high seas. Jim said the rest of the trip was even more eventful. After he had finished his watch in deteriorating weather, he went below to get some sleep. But he awoke to an unusual shuddering in the hull. He went back into the cockpit just as it was hit broadside by an 80-knot gust, which knocked the boat down almost flat. A leeward spreader was broken, weakening the mast support. Next day a gust slammed into the sails, breaking a second spreader and forcing a sail reduction to storm jib, reefed staysail, and mizzen. The weather settled down, and the boat bobbed along under self-steering gear. Weldy and Jim went below, looked at each other, and laughed.

Jim shares the respect of his siblings for his father. He credits Weldy's guidance through a challenging boyhood with his success in life. He recalls one incident at Resolute Bay that he will never forget, not so much for his own foolishness but for the reaction of his father afterwards.

"I went with some Eskimo friends on a hike of about 10 miles," he said. "My mother thought I had just gone to Resolute village, but we had gone out onto the shifting sea ice. It would be very easy to be carried out to sea and never seen again After I had been assumed missing, my father and brother David did an aerial search and spotted us. We got off the ice and returned home. My mother was pretty mad at me and wanted me severely disciplined. But Dad vetoed the idea of a spanking and said: 'I'm just glad to see you alive'."

~~~~~

## Honours and Awards

Welland W. Phipps was the recipient of extensive recognition of his outstanding contribution to the advancement of aviation in Canada. Principal among the awards was membership in the Order of Canada which was conferred at Rideau Hall, Ottawa, in 1976.

---

# The Chancellor and Principal Companion of the Order of Canada
## to *Welland W. Phipps*

### Greeting:

Whereas, with the approval of Her Majesty Queen Elizabeth the Second, Sovereign of the Order of Canada We have been pleased to appoint you to be a member of the Order of Canada

We do by these Presents appoint you to be a member of the said Order and authorize you to hold and enjoy the dignity of such appointment together with membership in the said Order and all privileges thereunto appertaining

Given at Rideau Hall in the City of Ottawa under the Seal of the Order of Canada this fourteenth day of January, 1976.

By the Chancellor's Command

*Esmond Butler*

Secretary General of the Order of Canada

## Trans-Canada McKee Trophy 1961

In recognition of his contribution to the cause of Canadian aviation through his development of landing gear which permits the exploitation of the short field characteristics of light aircraft operating from unprepared surfaces in the Arctic.

---

On the occasion of his acceptance of the Trans-Canada McKee Trophy, Weldy had the following comments before the meeting of the Air Transport Association of Canada:

> Mr. Chairman, Mr. Baldwin*, honoured guests, ladies and gentlemen:
>
> It is indeed with pride and humility that I accept the McKee Trophy today. In this modern day of supersonic flight and rocketry, a bush operation seems a little out of place. In fact I find my small contribution closely related to the first McKee Trophy winner. Thirty-five years ago "Doc" Oaks became the first recipient of the award for his flying endeavours which helped greatly in opening up the north country by air.
>
> In succeeding years other pilots were honoured for their work in the opening up the north. The names of Dickens, May, Tudhope and Phillips are familiar to you all, and I must say to be recognized with these men is a great privilege.
>
> Modern day advancements in bush flying can hardly be compared with the pioneering feats of the early pilots, but with the advantage of modern equipment I could not help feel that another step could be taken.

---

\* Deputy Minister of Transport

I won't go into the details here on the conversions of Cubs, Beavers and Otters and the development of flying techniques which made our job in the Arctic islands possible, but I would like to pay tribute to the many fine pilots who contributed so much to the success of our operations over the past five years.

The Arctic islands hold a great fascination for me and I am sure that their development could contribute immensely to the economic strength of Canada. In this regard I think the bush type operation will continue to play a major role.

We who operate in the Arctic have a clearer picture of the transport problems which exploration enterprises are faced with. No territory can be exploited or developed without transportation and no organization can undertake to explore a territory unless it is economically feasible.

The cost of air transport in the Arctic islands has always been a major deterrent to anyone considering an exploration venture.

The lack of a fixed base operation in the islands automatically tacks a punishing ferrying bill onto the overall cost and exploration parties invariably have to pay for a guaranteed number of flying hours in excess of their requirements.

This hindrance to the development of the Arctic islands can only be overcome by adequate and economical air transport being made available to exploration parties from a base in the islands and an adequate service to and from the islands.

A commercial operator cannot provide such a service unless he has a reasonable volume of business for seven or eight months of the year. The requirement for this volume exists at the present time but it is not available to the commercial operator. I am speaking of the flying carried out by the air force in supplying civil weather stations and other government departments.

I sincerely hope that the reorganized Air Transport Association will have some luck solving this problem. With this accomplished, the development of the Arctic islands would be greatly enhanced and the country may soon benefit from its vast resources.

I would like to thank the Department of National Defence for selecting me as McKee Trophy winner for 1961, and I thank the Air Transport Association for the honour of this occasion.

~~~~~

Weldy compares the standard Super Cub wheel to the 35-inch 'Phipps Special' he perfected in the fifties for use on the hostile landing terrain of the Arctic. It was one of many aircraft modifications he developed but undoubtedly the most enduring and productive.

Phipps collection

## Canada's Aviation Hall of Fame

The certificate, given at Edmonton, 31 December, 1993 is inscribed:

---

**In** *deserved recognition of conspicuous contributions to this nation through involvement with manned flight*

# W.W.Phipps

*has been named to rank with his peers as*

# MEMBER

*Canadian aviators have set permanent records...*

for unparalleled adventure... for enterprise... for pioneering achievements... for stubborn struggles against impossible odds... for unfailing courage... that have stood the test of time.

"the application of his aeronautical abilities in designing and perfecting the use of super-balloon aircraft tires, and his numerous flights into the high Arctic, have been of outstanding benefit to Canadian aviation."

## History of Canada's Aviation Hall of Fame

(as described at its inauguration 16 July 1994)

The Hall of Fame was designed as a "moveable feast", that all Canadian might learn of our aviation heritage; that the spirit of adventure might be awakened in the young and to preserve the personal history and enshrine the aeronautical achievements of those aviators serving Canada, living and dead, who have made the most outstanding contributions.

It was conceived by a selected group of legal historians, and given Federal Government sanction. To it were named Canada's six winners of the Victoria Cross in aerial combat, all of whom are dead, the 31 Companions of the Order of Icarus, Canada's most select aviation fraternity and the 38 winners of the McKee (Trans-Canada) Trophy, awarded since 1927 by the Department of National Defence, and now by the Canadian Aeronautics and Space Institute in Ottawa.

To this list were added Alexander Graham Bell and F. (Casey) Baldwin, the earliest of Canada's air pioneers.

Six winners of the McKee Trophy had previously been named to the order of Icarus, so the present membership in the Hall of Fame, including those named since the original list was compiled, stands at 79, of which 28 are deceased.

Nominations for membership are made to the Hall of Fame and pass before two, five-member committees, who look for "unselfish aviation contributions of major benefit to Canada, while engaged in aerial operations, and which have stood the test of time."

The directors were selected for their individual contributions to aviation, and the fact that between them these six men have explored flight from its near inception to the present, in every type of aircraft known to Canada; in peace and in three wars, including rotary, piston and jet aircraft, helicopters, gliders, parachutes and aerostats.

## The Order of Polaris

*By these presents be it known that*

# W. W. Phipps

*is hereby named to membership in the **Order of Polaris***

*In recognition of an outstanding role as a pioneer aviator in the Canadian North... for courage in navigating the uncharted wilderness... for dedication and perseverance in developing the Dominion's last frontiers, we the people of Canada's Yukon Territory are sincerely grateful and take pleasure in hereby bestowing this honour and its attendant decoration in perpetuity.*

*Given under the authority of the Commissioner of the Yukon Territory and the Yukon Legislative Council this 15th day of November, 1974.*

## Various other honours

### The P.E.I. FLYING ASSOCIATION
proudly presents
This Honourary Life Membership
to

*Welland (Weldy) Phipps*

"Whiskey Papa"
for his dedication innovation and
humanitarianism in pioneer aviation
of Canada's north
July 8, 1995

---

*Be it know by these present*
that

## W.W. Phipps
is hereby named to
**The Esteemed Brotherhood of Silver Wings
of the Northwest Territories**

This Honour is bestowed as a token in recognition of an outstanding
contribution made to the development of the Northwest Territories
through excellence in aviation. It shall be recognized
that because he met the challenge of Canada's North
the land was better charted, the earth yielded the secret of her riches
more readily, and understanding between men increased beyond
measure.
He shall therefore enjoy all the right and privileges attached to the
Brotherhood.
Granted under Authority and Seal of the Government
and Council of the Northwest Territories, Canada
in the Year 1974.

*S. M. Hodgson*
Commissioner of the Northwest Territories

## **Order of Flight**

Greetings:
Be it hereby known that

## *W.W. Phipps*

has been named to rank as

# COMPANION

with his peers in recognition of his superlative contributions as an aviation pioneer to this nation's growth and shall ever merit the highest esteem of the citizens of the great frontier City of Edmonton, in the Province of Alberta, in the Dominion of Canada. We, the people, pay tribute to his gallantry and exemplary dedication to the benefit of all our peoples, in the arena of manned flight. He shall henceforth be entitled to wear as a decoration the exclusive insignia of the Order and to place after his name the initials COF.

Give under authority of the Council of the City of Edmonton, in the Province of Alberta in the Year of Our Lord 1973.

*Ivor Dent,*

Mayor of the City of Edmonton

# Certificate of Merit

in recognition
of services rendered to the community
and to this organization the
**Saint Paul Area Chamber of Commerce**
presents this Certificate to

*Weldy Phipps*

---

(given by St. Paul Minnesota CofC after the Plaisted Expedition)

## A final word...

Weldy hired me to fly for Spartan back in 1951. He was my boss. When I joined COPA (Canadian Owners and Pilots Association) in 1957, I succeeded in getting Weldy on the board of directors. He became my boss again.

All I can say is that Weldy is one of the few people I have met in the 50 years that I've been making a living in aviation, in whom I would have complete confidence that whatever job he tackled would be accomplished.

He was just plain smart. He was a problem solver. Weldy was the type of guy who must be put in charge of an operation with no strings attached if it is to be highly successful. He needed the freedom to improvise in a most unorthodox way. He was not a person to stand on formality or abide by precedent.

He was not only good in technical and mechanical matters but he was good at handling people in a rather quiet and unassuming way. He could size people up within two minutes of talking with them and judge their capability. He expected all employees to show the same enthusiasm and dedication to their work as he showed. He was a slave driver, to use a common expression, but an admired slave driver who was well respected. There was no such thing as a 5-day week or 8-hour day when he was in charge. We worked 365 days a year, if necessary, and often a 16-hour day -- in fact, right around the clock on some occasions. Weldy and C.D. Howe would have got along just fine.

We spent much time together but we were always on the go. I never knew Weldy to relax and if he was having a beer with the boys, you can be assured that the conversation was about flying and some aspects of his work. He had an enormous amount of PMA.

He did well at Rimouski Airlines, at Spartan and at Atlas. But he is the type of person who could have done well running General Motors or the Royal Bank of Canada if given the opportunity.

--Bill Peppler